Maple V
Language Reference Manual

Bruce W. Char   Keith O. Geddes   Gaston H. Gonnet
Benton L. Leong   Michael B. Monagan   Stephen M. Watt

Maple V
Language Reference Manual

Springer-Verlag
New York Berlin Heidelberg London Paris
Tokyo Hong Kong Barcelona Budapest

Bruce W. Char
Department of Mathematics
  and Computer Science
Drexel University
Philadelphia, PA 19104
U.S.A.

Keith O. Geddes
Department of Computer Science
University of Waterloo
Waterloo, ON
Canada N2L 3G1

Gaston H. Gonnet
Department Informatik
ETH Zentrum
8092 Zürich
Switzerland

Benton Leong
Symbolic Computation Group
University of Waterloo
Waterloo, ON
Canada N2L 3G1

Michael B. Monagan
Department Informatik
ETH Zentrum
8092 Zürich
Switzerland

Stephen M. Watt
IBM Thomas J. Watson
  Research Center
P.O. Box 128
Yorktown Heights, NY 10598
U.S.A.

Library of Congress Cataloging-in-Publication Data
Maple V Language reference manual /
   Bruce Char . . . [et al.] 1st ed.
      p.   cm.
   Includes bibliographical references and index.
   1. Maple (Computer program)   I. Char, Bruce W.
   QA155.7E4M36 1991
   510.′285′5369dc20   91-20181

Printed on acid-free paper.

Photocomposed copy prepared from the authors' TeX files.
Printed and bound by R.R. Donnelley & Sons, Harrisonburg, VA.
Printed in the United States of America.

9 8 7 6 5 4 3 2 1

ISBN 0-387-97622-1 Springer-Verlag New York Berlin Heidelberg
ISBN 3-540-97622-1 Springer-Verlag Berlin Heidelberg New York

Preface

The design and implementation of the Maple system is an on-going project of the Symbolic Computation Group at the University of Waterloo in Ontario, Canada. This manual corresponds with version V (roman numeral five) of the Maple system. The on-line *help* subsystem can be invoked from within a Maple session to view documentation on specific topics. In particular, the command `?updates` points the user to documentation updates for each new version of Maple.

The Maple project was first conceived in the autumn of 1980 growing out of discussions on the state of symbolic computation at the University of Waterloo. The authors wish to acknowledge many fruitful discussions with colleagues at the University of Waterloo, particularly Morven Gentleman, Michael Malcolm, and Frank Tompa. It was recognized in these discussions that none of the locally-available systems for symbolic computation provided the facilities that should be expected for symbolic computation in modern computing environments. We concluded that since the basic design decisions for the then-current symbolic systems such as ALTRAN, CAMAL, REDUCE, and MACSYMA were based on 1960's computing technology, it would be wise to design a new system from scratch taking advantage of the software engineering technology which had become available since then, as well as drawing from the lessons of experience.

Maple's basic features (e.g. elementary data structures, input/output, arithmetic with numbers, and elementary simplification) are coded in a systems programming language for efficiency. For users, there is a high-level language with a modern syntax more suitable for describing algebraic algorithms. An important property of Maple is that most of the algebraic facilities in the system are implemented using the high-level user language. The basic system, or kernel, is sufficiently compact and efficient to be practical for use in a shared environment or on personal computers with as little as two megabytes of main memory. Library functions are loaded into the system as required, adding to the system facilities such as polynomial factorization, equation solving, indefinite integration, and matrix manipulation functions. The modularity of this design allows users to consume computer resources proportional to the algebraic facilities actually being used.

The system kernel is written in macros which can be translated by a locally-developed macro processor (called Margay) into versions of the kernel in the C programming language for various operating systems. Operating systems currently supporting a Maple implementation include UNIX (and various UNIX-like systems), 386 DOS, Macintosh Finder, DEC VMS, IBM VM/CMS and Amiga DOS.

Acknowledgments

This manual has benefited from comments given to us by students, faculty, and staff at the University of Waterloo, and by users at sites throughout the world. These "friends of Maple" are too numerous to list here. We hope they will feel that their assistance has been acknowledged when they find their suggestions incorporated into the present edition. For their assistance in the production of this manual we would like to thank Kate Atherley, Greg Fee, Krishna Gopinathan, Blair Madore, Liyuan Qiao, Marc Rybowicz, and Katy Simonsen. We also thank Stefan Vorkoetter of Waterloo Maple Software for his contribution of the appendix on using Maple under DOS.

The Maple project has received support from various sources, including the Academic Development Fund of the University of Waterloo, the Natural Sciences and Engineering Research Council of Canada, Digital Equipment Corporation, the Sloan Foundation, and the Information Technology Research Centre of Ontario.

Contents

1
Introduction

Maple is a mathematical manipulation language. (The name can be said to be derived from some combination of the letters in the preceding phrase, but in fact it was simply chosen as a name with a Canadian identity.) The type of computation provided by Maple is known by various other names such as *algebraic manipulation*, *symbolic computation*, or *computer algebra*. A basic feature of such a language is the ability to, explicitly or implicitly, leave the elements of a computation unevaluated. A corresponding feature is the ability to perform simplification of expressions and other transformations involving unevaluated elements. For example,

```
> f:= x^3 - y^3 ;
```
$$f := x^3 - y^3$$

```
> factor(f);
```
$$(x - y)(x^2 + x y + y^2)$$

Readers familiar with traditional programming languages, such as Fortran, Pascal, C, LISP, or APL, will notice that symbols (identifiers) in the above example are used for two different purposes. The symbol f is used as a variable, it is assigned a value. The symbols x and y are used as unknowns, as yet they have no value.

In Maple, statements are normally evaluated as far as possible in the current environment. For example, the statement

```
a := 1;
```

assigns the value 1 to the name a. If this statement is later followed by the statement

```
x := a + b;
```

then the value 1+b is assigned to the name x. Next if the assignments

```
b := -1;    f := sin(x);
```

are performed then x evaluates to 0 and the value 0 is assigned to the name f. (Note that sin(0) is automatically "simplified" to 0.) Finally if we now perform the assignments

```
b := 0;    g := sin(x);
```

then x evaluates to 1 and the value sin(1) is assigned to the name g. (Note that sin(1) cannot be further evaluated or simplified in a symbolic context, but there is a facility to evaluate an expression into floating-point form which will yield the decimal expansion of `sin(1)` to a number of digits controlled by the user.)

As each complete statement is entered by a user, it is evaluated and the results are printed on the output device, usually the user's terminal. The printing of expressions is normally presented in a two-dimensional, multi-line format designed to be as readable as possible on standard computer terminals.   There are several ways in which the user can alter the display of expressions. If the interface variable `prettyprint` is set to `false` via the command `interface(prettyprint=false)` then expressions are printed in a one-dimensional line-printing mode that is more suitable for machine input than for human readability. The default value of `prettyprint` is `true`, resulting in a two-dimensional display which is centered in the case of a single-line expression and which breaks the expression onto several lines of two-dimensional output as required.

The global variable `printlevel` determines how much output will be printed. Setting this variable to `-1` prevents the printing of any results. The default value of `printlevel` is `1`, and in general, higher integer values cause more intermediate steps in a computation to be displayed. (See Chapter 11 for more details.) Finally, the silent statement separator ':', when used to terminate a statement at the interactive level, prevents the results of the statement from being displayed. (The normal statement separator is ';'.)

There is an on-line *help* facility in the Maple system. To start using it, enter the command ? and some information about the help facility will be displayed at the terminal. Some examples of particular requests for help pages are:

```
?index
?gcd
?if
?linalg[eigenvals]
```

The following pages contain the transcript of an interactive session with Maple, similar to what one would see at a terminal. Lines beginning with '>' are user input lines. Centered lines are system responses.

A Maple session is typically initiated by the command `maple` . When a session is initiated, the maple leaf logo is displayed and a prompt character, such as '>', appears. (This prompt character may be different on some systems.)

```
      |\^/|      MAPLE V
 ._|\|   |/|_.  Copyright (c) 1981-1990 by the University of Waterloo
  \  MAPLE  /   All rights reserved. MAPLE is a registered trademark of
  <____ ____>   Waterloo Maple Software
       |        Type ? for help.
 >
```

1.1   Some General Examples

1.1.1   Polynomials

Maple supports univariate and multivariate polynomials in both expanded and unexpanded form.

```
> (x + 1)^4 * (x + 2)^2;
```

$$(x + 1)^4 \ (x + 2)^2$$

```
> expand(");
```

$$x^6 + 8 x^5 + 26 x^4 + 44 x^3 + 41 x^2 + 20 x + 4$$

```
> factor(");
```

$$(x + 1)^4 \ (x + 2)^2$$

The double-quote symbol '"' refers to the latest expression. (Similarly the '""' command refers to the second last expression, and '"""' refers to the third last expression.)

Notice that every Maple statement ends with a semicolon. The exponentiation operator is "^" (Maple understands "**" to be a synonym for "^"). Multiplication is denoted by "*", and division by "/". The assignment operator is ":=" as in the programming language Pascal.

```
> a := expand((x*y/2 - y^2/3) * (x - y) * (3*x + y));
```

$$a := 3/2 \ x^3 \ y - 2 \ x^2 \ y^2 + 1/6 \ x \ y^3 + 1/3 \ y^4$$

Maple does not automatically transform rational expressions into a standard (canonical) form.

```
> a / (x^3 - x^2*y - x*y + y^2);
```

$$\frac{3/2 \ x^3 \ y - 2 \ x^2 \ y^2 + 1/6 \ x \ y^3 + 1/3 \ y^4}{x^3 - x^2 \ y - x \ y + y^2}$$

However, facilities are available for simplification on demand.

```
> normal(") ;
```

$$1/6 \ \frac{(9 \ x^2 - 3 \ x \ y - 2 \ y^2) \ y}{x^2 - y}$$

There are explicit functions for greatest common divisor and least common multiple computations with polynomials. Ending a command with a ":" causes Maple to compute the result without printing it.

```
> gcd(x^3 + 1, x^2 + 3*x + 2) ;
```

$$x + 1$$

```
> p := 55*x^2*y + 44*x*y - 15*x*y^2 - 12*y^2:
```

```
> q := 77*x^2*y - 22*x^2 - 21*x*y^2 + 6*x*y:
> gcd(p, q) ;
```
$$11\ x\ -\ 3\ y$$

```
> lcm(15*(x - 5)*y, 9*(x^2 - 10*x + 25)) ;
```
$$45\ x^2\ y\ -\ 450\ x\ y\ +\ 1125\ y$$

1.1.2 Factoring

Maple knows how to factor polynomials over various domains, including the integers, finite fields, and algebraic number fields.

```
> factor( x^4-2 ) ;
```
$$x^4\ -\ 2$$

```
> factor( x^4-2, sqrt(2)) ;
```
$$(x^2\ +\ 2^{1/2}\ )\ (x^2\ -\ 2^{1/2}\ )$$

The second parameter to factor specifies the algebraic extension. This parameter can be specified as a radical, as above, or a RootOf. RootOf(x^4-2) refers to *any* root of $x^4 - 2$. The **alias** command is used to define the symbol **alpha** as an abbreviation to be used for input and output. This helps to keep the output readable.

```
> alias(alpha=RootOf(y^2-2)):
> factor(x^4-2,alpha) ;
```
$$(x^2\ +\ alpha)\ (x^2\ -\ alpha)$$

In the following example, we find that although $x^6 + x^5 + x^4 + x^3 + 1$ is irreducible over the integers, it is factorable over the integers modulo 2. Note the use of "Factor" rather than "factor". This prevents Maple from prematurely attempting to invoke the **factor** command, which works over the integers.

```
> f := x^6+x^5+x^4+x^3+1 ;
```
$$f := x^6\ +\ x^5\ +\ x^4\ +\ x^3\ +\ 1$$

```
> Factor(f) mod 2 ;
```
$$(x^4\ +\ x\ +\ 1)\ (x^2\ +\ x\ +\ 1)$$

1.1.3 Help

Maple has a comprehensive on-line help facility. Help for Maple functions is available through the command "?function". Help on the help command itself is available via "?". Let's see what Maple says about the "**factor**" function.

```
> ?factor
```

FUNCTION: factor - factor a multivariate polynomial

CALLING SEQUENCES:
   factor(a); or factor(a,K);

PARAMETERS:
   a - an expression
   K - an algebraic extension

SYNOPSIS:
- The function factor computes the factorization of a multivariate polynomial
  with integer, rational or algebraic number coefficients.

- If the input is a rational function, then a is first ``normalized'' (see normal) and the numerator and denominator of the resulting expression are then
  factored. This provides a ``fully-factored form'' which can be used to simplify in the same way the normal function is used. However, it is more
  expensive to compute than normal.

- If the input a is a list, set, equation, range, series, relation, or function, then factor is applied recursively to the components of a.

- The call factor(a,K) factors a over the algebraic number field defined by K.
  K must be a single RootOf or a list or set of RootOf's or a single radical or
  a list or set of radicals.

- If the 2nd argument K is not given, the polynomial is factored over the
  rationals. Note that any integer content (see first example below) is not
  factored.

EXAMPLES:
```
> factor(6*x^2+18*x-24);
```
$$6 \ (x + 4) \ (x - 1)$$

```
> factor({x^3+y^3 = 1});
```
$$\{(x + y) \ (x^2 - x \ y + y^2) = 1\}$$

```
> factor((x^3-y^3)/(x^4-y^4));
```
$$\frac{x^2 + x \ y + y^2}{(x + y) \ (x^2 + y^2)}$$

SEE ALSO: ifactor, Factor, factors, collect, galois, irreduc, roots

1.2 Numbers

Of course, Maple can also do arithmetic. However, unlike your calculator, rational arithmetic is exact

```
> 1/2 + 1/3 + 2/7 ;
                                    47
                                   ----
                                    42
```

and calculations are performed using as many digits as necessary.

```
> p := 1153 * (3^58 + 5^40) / (29! - 7^36) ;

                         1380507571497552721407199431 4
                  p := ------------------------------
                         536863065731140371445324598 3
```

Constants in Maple may be approximated by floating-point numbers with the user having control over the number of digits carried via the global variable "**Digits**"(its default value is 10). The "**evalf**" function causes evaluation to a floating-point number.

```
> evalf(p) ;
                              2.571433313
```

Floating-point arithmetic takes place automatically when floating-point numbers are present.

```
> 1/2.0 + 1/3.0 + 2/7.0 ;
                              1.119047619
```

Trigonometric functions use radians.

```
> h := tan(3*Pi/7) ;
                           h := tan(3/7 Pi)

> evalf(h) ;
                              4.381286277

> Digits := 40 ;
                             Digits := 40

> evalf(h) ;
                  4.381286267534823072404689085032695444160
> Digits := 10 ;
                             Digits := 10
```

Maple knows about many other special functions besides trigonometric functions. If the argument to these functions is a floating-point number, evaluation occurs automatically.

```
> erf(3) ;
                                erf(3)

> evalf(") ;
                              .9999779095

> Zeta(2.2) ;
                              1.490543257
```

Maple can factor integers and test for primality.

```
> p := 2^(2^6)+1 ;
                        p := 18446744073709551617
```

```
> isprime(p) ;
                                false
```

```
> ifactor(p) ;
                        (67280421310721) (274177)
```

Maple knows about many kinds of special numbers such as the binomial coefficients, Bernoulli,[1]
Euler, Fibonacci, Bell, and Stirling numbers.

```
> bernoulli(4) ;
                                -1/30
```

```
> seq(binomial(10,k), k=0..10) ;
                1, 10, 45, 120, 210, 252, 210, 120, 45, 10, 1
```

1.3 Examples from Calculus

1.3.1 Differentiation

Differentiation in Maple is performed with the "diff" command. The first argument to diff is
the expression to be differentiated, the second argument is the variable with respect to which to
differentiate.

```
> diff(sin(x) * cos(x), x) ;
                                2         2
                            cos(x)  - sin(x)
```

```
> diff(sin(x) * x^(x^x), x) ;

              x                 x /                         x \
           (x )             (x ) | x                      x |
    cos(x) x       + sin(x) x     |x  (ln(x) + 1) ln(x) + ----|
                                 \                         x /
```

```
> diff(g(x), x) ;
                                 d
                                ---- g(x)
                                 dx
```

---

[1] Also the Bernoulli and Euler polynomials $B_n(x)$ and $E_n(x)$.

```
> diff(erf(x), x) ;
```

$$2 \, \frac{\exp(- x^2)}{Pi^{1/2}}$$

What does the error function **erf(x)** look like? Let's plot it, together with its derivative (see Figure 1.1).

```
> plot( {erf(x),diff(erf(x),x)}, x=-5..5 );
```

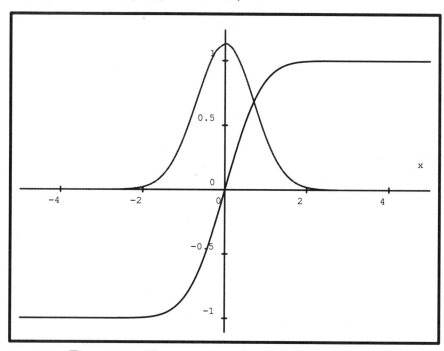

Figure 1.1. The error function and its derivative

Sometimes you want to find the third derivative without finding the first and second.

```
> f := x^9+3*x^7-x^5+5*x^3+1 ;
```

$$f := x^9 + 3 x^7 - x^5 + 5 x^3 + 1$$

```
> diff(f,x,x,x) ;
```

$$504 x^6 + 630 x^4 - 60 x^2 + 30$$

Of course we have an abbreviation to make this easier.

```
> x$6 ;
```

$$x, \, x, \, x, \, x, \, x, \, x$$

```
> diff(f, x$6) ;
```

$$60480 x^3 + 15120 x$$

1.3.2 Mathematical Functions

Math functions of one or more variables can easily be defined in a very natural way. These functions can be evaluated, again in a natural way, at numeric or symbolic values.

```
> f := x -> sin(x)/x;
```

$$f := x \rightarrow \frac{\sin(x)}{x}$$

```
> f(2.0) ;
```

$$.4546487134$$

```
> f(x) ;
```

$$\frac{\sin(x)}{x}$$

```
> g := (x,y) -> (x^2-y^2)/(x^2+y^2) ;
```

$$g := (x,y) \rightarrow \frac{x^2 - y^2}{x^2 + y^2}$$

```
> g(1,2) ;
```

$$-3/5$$

```
> g(1,x) ;
```

$$\frac{1 - x^2}{1 + x^2}$$

They can be also be graphed (see Figure 1.2).

```
> plot3d(g, -1..1, -1..1) ;
```

Functions can be differentiated with the *D* operator. The result is another function.

```
> D(sin+cos) ;
```

$$\cos - \sin$$

```
> D(f) ;
```

$$x \rightarrow \frac{\cos(x)}{x} - \frac{\sin(x)}{x^2}$$

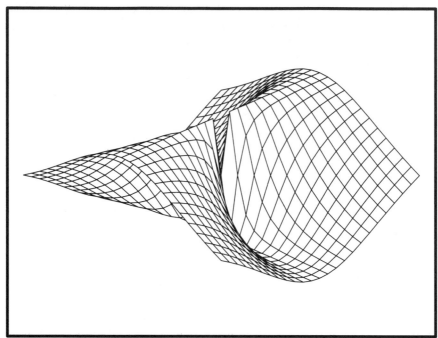

Figure 1.2. A surface plot of $\frac{x^2-y^2}{x^2+y^2}$

The D operator can also be used to find partial derivatives. $D[i](h)$ refers to the partial derivative of h with respect to the ith argument.

```
> h := (x,y) -> x^5 - x^2*y - y^2 ;
                              5    2     2
                 h := (x,y) -> x  - x  y - y
> D[2](h);
                                 2
                    (x,y) -> - x  - 2 y
> D[1,2](h);
                    (x,y) -> - 2 x
```

1.3.3 Integration

An important facility in Maple is analytic integration. This is performed with the int command, whose syntax is similar to that of the diff command.

```
> f := (x^3+9*x^2+28*x+27)/((x+3)^3*x) ;
```

$$f := \frac{x^3 + 9x^2 + 28x + 27}{(x+3)^3 \, x}$$

```
> int(f, x) ;
```

$$- \frac{1}{2(x+3)^2} + \ln(x)$$

```
> int(f, x = 1..2) ;
```

$$9/800 + \ln(2)$$

```
> int(x^5 / (43 - x^2)^(19/2), x) ;
```

$$1/13 \; \frac{x^4}{(43 - x^2)^{17/2}} + \frac{172}{195 (43 - x^2)^{15/2}} - \frac{7396}{221 (43 - x^2)^{17/2}}$$

```
> int(ln(t)^2*t^(1/2)*exp(-t), t = 0..infinity) ;
```

$$1/4 \; Pi^{5/2} - 2 \; Pi^{1/2} \; gamma - 4 \; Pi^{1/2} \; \ln(2) + 1/2 \; Pi^{1/2} \; gamma^2$$

$$+ 2 \; Pi^{1/2} \; gamma \; \ln(2) + 2 \; Pi^{1/2} \; \ln(2)^2$$

We can find a floating-point approximation of this quantity. [2]

```
> evalf(") ;
```

$$.829626907$$

When Maple cannot find a closed form solution for an integral (definite or indefinite), it returns the input command which gets pretty-printed using a stylized integral sign. For example

```
> int(exp(-t)/(1+t^(3/2)), t=0..infinity) ;
```

$$\int_{0}^{\infty} \frac{\exp(-t)}{1 + t^{3/2}} \, dt$$

The resulting expression can be manipulated with other Maple tools. In this example, since the result is a constant we can use **evalf** to find a numerical approximation for it (i.e. numerical integration is performed).

---

[2]Note: **gamma** is Euler's constant $\gamma = 0.5772156649....$

```
> evalf(") ;
```

$$.6130734052$$

In the following example we compute the Taylor series of the resulting function. We shall present more examples of Maple's ability to calculate series in a later section.

```
> int(exp(x^3), x) ;
```

$$\int exp(x^3)\ dx$$

```
> taylor(", x=0, 15) ;
```

$$x + 1/4\ x^4 + 1/14\ x^7 + 1/60\ x^{10} + 1/312\ x^{13} + O(x^{16})$$

Some integrals do have rather unwieldy closed forms.

```
> x/(x^3-x^2+1) ;
```

$$\frac{x}{x^3 - x^2 + 1}$$

```
> int(",x=0..1) ;
```

$$\left(\sum_{\_R = \%1} \_R\ \ln(4/3 + 23/3\ \_R^2)\right) - \left(\sum_{\_R = \%1} \_R\ \ln(23/3\ \_R^2 + 1/3)\right)$$

$$\%1 := RootOf(23\ \_Z^3 + 3\ \_Z + 1)$$

1.3.4 Summation

Maple, as seen by the result of the above integral, can also handle summation. Summation over the roots of a polynomial can be evaluated numerically. To do this Maple approximates the roots numerically and then sums over them. For example, numerical evaluation of the above integration result yields:

```
> evalf(");
```

$$.556825463$$

Maple can also handle more conventional summations over both numeric and symbolic ranges. The syntax is equivalent to the **int** command.

```
> sum(i, i=0..n-1) ;
```

$$1/2\ n^2 - 1/2\ n$$

```
> sum(i^2, i=1..9876543210) ;
```
$$321139442946604922593984029585$$

```
> sum(binomial(n, i), i=0..n) ;
```
$$2^n$$

```
> sum(i^4 * 7^i, i=1..n-1) ;
```
$$-\frac{91}{54}\, 7^n\, n + \frac{70}{81}\, 7^n + 14/9\, 7^n\, n^2 - 7/9\, 7^n\, n^3 + 1/6\, n^4\, 7^n - \frac{70}{81}$$

```
> sum((i - 1) / (i + 1), i=1..k) ;
```
$$k + 2 - 2\, \mathrm{Psi}(k + 2) - 2\, \mathrm{gamma}$$

```
> sum(i^4 * 4^i / binomial(2*i,i), i=0..n-1) ;
```
$$4/693\, \frac{(2n - 1)\, 4^{(n - 1)}\, (-6 + 26n + 60 n^2 - 140 n^3 + 63 n^4)}{\mathrm{binomial}(2n, n)} - 2/231$$

```
> sum(1/i^2 + 1/i^3, i=1..infinity) ;
```
$$1/6\, \mathrm{Pi}^2 + \mathrm{Zeta}(3)$$

1.3.5 Limits

There is a limit function to compute the limiting value of an expression as a specified variable approaches a specified value.

```
> r := (sin(x)-x)/x^3 ;
```
$$r := \frac{\sin(x) - x}{x^3}$$

```
> limit(r, x=0) ;
```
$$-1/6$$

```
> limit(r, x=infinity) ;
```
$$0$$

```
> limit(Psi(2*exp(x))-x, x=infinity) ;
```
$$\ln(2)$$

1.3.6 Series

Maple has facilities for computing series expansions of expressions.

```
> series(ln(1+z), z=0) ;
```

$$z - 1/2\ z^2 + 1/3\ z^3 - 1/4\ z^4 + 1/5\ z^5 + O(z^6)$$

```
> series(erf(x), x=0) ;
```

$$\frac{2}{\text{Pi}^{1/2}}\ x - \frac{2}{3\ \text{Pi}^{1/2}}\ x^3 + \frac{1}{5\ \text{Pi}^{1/2}}\ x^5 + O(x^6)$$

The $O(\ )$ term indicates the order of truncation of the series. The order of truncation can be controlled by the user; the default value is order 6.

```
> series(exp(-x^2)*(1-erf(x)), x=0, 4 ) ;
```

$$1 - \frac{2}{\text{Pi}^{1/2}}\ x - x^2 + \frac{8}{3\ \text{Pi}^{1/2}}\ x^3 + O(x^4)$$

The result returned by "**series**" isn't necessarily a Taylor series. In the following example the result returned is a Laurent series.

```
> series(GAMMA(x),x=0,3) ;
```

$$x^{-1} - \text{gamma} + (1/12\ \text{Pi}^2 + 1/2\ \text{gamma}^2)\ x$$

$$+ (-\ 1/3\ \text{Zeta}(3) - 1/12\ \text{Pi}^2\ \text{gamma} - 1/6\ \text{gamma}^3)\ x^2 + O(x^3)$$

```
> evalf(");
```

$$1.\ x^{-1} - .5772156649 + .9890559955\ x - .9074790760\ x^2 + O(x^3)$$

Here are some examples using Bessel functions. Note that instead of using the longer names **BesselJ** and **BesselK** that Maple uses for the Bessel functions $J_v(x)$ and $K_v(x)$, we have defined J and K to be abbreviations by using the **alias** command.

```
> alias(J=BesselJ, K=BesselK):
> series(J(0,x), x=0, 10) ;
```

$$1 - 1/4\ x^2 + 1/64\ x^4 - 1/2304\ x^6 + 1/147456\ x^8 + O(x^{10})$$

The next series command results in a "generalized series" rather than a pure Laurent series. Note the $\ln(x)$ term in the coefficient of x^3.

```
> series(K(3,x), x=0, 5) ;
```

$$8 x^{-3} - x^{-1} + 1/8 x + \left(- 1/48 \ln(2) + 1/48 \ln(x) + 1/48 \ \mathrm{gamma} - \frac{11}{576}\right) x^3 + O(x^5)$$

Maple knows how to compute with asymptotic series as well.

```
> f := n * (n + 1) / (2*n - 3) ;
```

$$f := \frac{n \ (n + 1)}{2 \ n - 3}$$

```
> asympt(f, n) ;
```

$$1/2 \ n + 5/4 + \frac{15}{8 \ n} + \frac{45}{16 \ n^2} + \frac{135}{32 \ n^3} + \frac{405}{64 \ n^4} + \frac{1215}{128 \ n^5} + O\left(\frac{1}{n^6}\right)$$

```
> asympt(erf(x), x) ;
```

$$1 + \frac{- \dfrac{1}{\mathrm{Pi}^{1/2} \ x} + \dfrac{1}{2 \ \mathrm{Pi}^{1/2} \ x^3} - \dfrac{3}{4 \ \mathrm{Pi}^{1/2} \ x^5} + O\left(\dfrac{1}{x^7}\right)}{\exp(x^2)}$$

The output from the series command is an approximation of some function. This can be used to construct further approximations, such as Padé [3] approximations. In the following example we compare $\sin(x)$ with its Padé approximation by plotting the result (see Figure 1.3).

```
> series(sin(x), x=0,10) ;
```

$$x - 1/6 \ x^3 + 1/120 \ x^5 - 1/5040 \ x^7 + 1/362880 \ x^9 + O(x^{10})$$

```
> approx := convert(", ratpoly) ;
```

$$\mathrm{approx} := \frac{x - \dfrac{53}{396} x^3 + \dfrac{551}{166320} x^5}{1 + \dfrac{13}{396} x^2 + 5/11088 \ x^4}$$

```
> plot({approx, sin(x)}, x=-3*Pi/2...3*Pi/2) ;
```

---

[3] Continued-fraction and Chebyshev series approximations are also available.

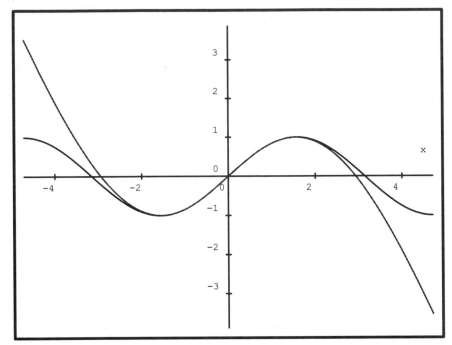

Figure 1.3. A plot of $\sin(x)$ and a rational approximation of $\sin(x)$

1.4   Data Structures

Maple is a programming language as well as a mathematical system. As such it supports a variety of data structures such as arrays and tables as well as mathematical objects like polynomials and series. Sets and lists are two other basic data structures that we will use in the next sections. They are constructed from sequences, that is, expressions separated by commas.

```
> s := sin, cos, tan ;
                         s := sin, cos, tan
```

Sequences can be generated using the **seq** command.

```
> seq(i^2, i=1..10) ;
                1, 4, 9, 16, 25, 36, 49, 64, 81, 100
```

1.4.1   Sets and Lists

Sets are represented using braces and the set operators are "**union**", "**intersect**", "**minus**" (for set difference), and "**member**" (for set membership). Note that the order of the elements in a set is arbitrary (in the mathematical concept of a set), and the Maple system chooses a specific ordering based on the internal addresses of the expressions.

```
> a := {exp, sin, cos} ;
```

```
                              a := {cos, exp, sin}

> b := {s} ;
                              b := {cos, sin, tan}

> a union b ;
                              {cos, exp, sin, tan}

> a minus b ;
                                    {exp}

> a intersect b ;
                                 {cos, sin}

> member(sin, b) ;
                                    true
```

Lists, unlike sets, retain the user-specified order and multiplicity. They are represented using square brackets.

```
> c := [1, 2, 3, 2, 1] ;
                              c := [1, 2, 3, 2, 1]
```

Selection of elements from sets and lists (and more generally, selection of operands from any expression) is accomplished using the "op" function. op(i,expr) yields the ith operand. Also, nops(expr) yields the number of operands in expr.

```
> nops(c) ;
                                     5

> op(2, c) ;
                                     2
```

An alternative selection operation is as follows.

```
> c[2] ;
                                     2
```

List concatenation is performed using op.

```
> d:=[2, 5, 46];
                              d := [2, 5, 46]

> e:=[op(c), op(d)];
                        e := [1, 2, 3, 2, 1, 2, 5, 46]
```

Functions may be applied to the elements of a list or set using the map command.

```
> map(x -> x^2, c) ;

                                [1, 4, 9, 4, 1]

> convert(b, list);

                                [cos, sin, tan]

> map(D, ") ;
```

$$[- \sin, \cos, 1 + \tan^2]$$

And one can iterate over the elements of a set or list in a natural way in a loop.

```
> for f in b do diff(f(x),x) od ;
```

$$- \sin(x)$$

$$\cos(x)$$

$$1 + \tan(x)^2$$

1.4.2 Arrays and Tables

Maple has the standard "**array**" data structure found in most programming languages. Arrays are created using the **array** command which defines the dimensions of the array. The syntax for selecting an element is similar to other programming languages. For example, the following creates a one-dimensional array of data and then selects the second element.

```
> u := array(1..4, [1.32, 4.57, 9.87, 11.4]) ;
                u := [ 1.32, 4.57, 9.87, 11.4 ]

> u[2];
```
$$4.57$$

Entries of an array can also be assigned in a for loop. Here we create a two-dimensional array.

```
> A := array(1..3,1..3):
> for i to 3 do for j to 3 do A[i,j] := i^2-j^2 od od:
> print(A);
```
$$\begin{bmatrix} 0 & -3 & -8 \\ 3 & 0 & -5 \\ 8 & 5 & 0 \end{bmatrix}$$

In addition to arrays, Maple also has a "**table**" facility. Unlike arrays, tables can be defined ad-hoc and indices can be any value, not just integers. Tables are created as soon as you assign something to an indexed variable which is not already an array or table.

```
> Height[Joe] := 175 * cm ;
```
$$\text{Height[Joe]} := 175 \text{ cm}$$

```
> Height[Fred] := 185 * cm ;
```
$$\text{Height[Fred]} := 185 \text{ cm}$$

```
> Height[average] := (Height[Joe] + Height[Fred])/2 ;
```
$$\text{Height[average]} := 180 \text{ cm}$$

Of course you can define a variable to be a table. You can also specify all the entries directly and there are functions for accessing the table indices and entries.

```
> COLORS := table([ red=(rot,rouge), blue=(blau,bleu) ]) ;
```

```
COLORS := table([
                blue = (blau, bleu)
                red = (rot, rouge)
         ])
```

```
> for c in entries(COLORS) do print(c) od;
                [blau, bleu]

                [rot, rouge]
```

1.5   Examples from Linear Algebra

The linear algebra package "linalg" allows standard matrix manipulation as well as many other functions. To use a function in the linalg package we could say "linalg[functionname]". To avoid using the long names for linalg functions we first tell Maple we want to use the linear algebra package.

```
> with(linalg):
Warning: new definition for    norm
Warning: new definition for    trace
```

The "with" function sets up definitions of the functions in a package such as "linalg". After the "with", we can use "det" instead of "linalg[det]". Had we ended the "with" command with ";" instead of ":" we would have seen a complete list of all the functions in the "linalg" package, which is too long to include here. Naming conflicts are always reported to ensure the user is informed.

In Maple, vectors are represented as one dimensional arrays and may be created using the vector or array functions. We use the angle command to compute the angle between two vectors.

```
> u := vector([1,2,3]) ;
                u := [ 1, 2, 3 ]

> v := vector([0,0,1]) ;
                v := [ 0, 0, 1 ]

> angle(u,v) ;
                          1/2
            arccos(3/14 14    )
```

Matrices are represented as two dimensional arrays. The elements of a matrix can be specified row-by-row in the matrix function, and they are displayed in a two-dimensional format. Note the use of the doubly nested lists.

```
> a := matrix([[x,y,z],[y,x,y],[z,y,x]]);
                [ x  y  z ]
                [         ]
           a := [ y  x  y ]
                [         ]
                [ z  y  x ]
```

```
> det(a) ;
```

$$x^3 - 2 x^2 y + 2 z^2 y - z^2 x$$

```
> factor(") ;
```

$$(x - z) (x^2 + z x - 2 y^2)$$

Alternatively, matrix entries may be created in a single list if you specify the row and column dimensions.

```
> b := matrix(3,3,[1,2,3,-2,1,-2,-3,2,1]) ;
```

$$b := \begin{bmatrix} 1 & 2 & 3 \\ -2 & 1 & -2 \\ -3 & 2 & 1 \end{bmatrix}$$

```
> inverse(b) ;
```

$$\begin{bmatrix} 5/18 & 2/9 & -7/18 \\ 4/9 & 5/9 & -2/9 \\ -1/18 & -4/9 & 5/18 \end{bmatrix}$$

Let us find the eigenvalues[4] of the matrix b using the builtin command **eigenvals**.

```
> eigenvals(b) ;
```

$$1, \ 1 + I \ 17^{1/2}, \ 1 - I \ 17^{1/2}$$

Now let's do the same computation step-by-step by forming the characteristic polynomial.

```
> charmat(b, lambda) ;
```

$$\begin{bmatrix} lambda - 1 & -2 & -3 \\ 2 & lambda - 1 & 2 \\ 3 & -2 & lambda - 1 \end{bmatrix}$$

```
> det(") ;
```

$$lambda^3 - 3 \ lambda^2 + 20 \ lambda - 18$$

```
> solve(", lambda) ;
```

$$1, \ 1 + I \ 17^{1/2}, \ 1 - I \ 17^{1/2}$$

Numerical computation of eigenvalues can also be performed.

```
> evalf( Eigenvals(b) );
```

$$[\ 1.000000000 + 4.123105624 \ I, \ 1.000000000 - 4.123105624 \ I, \ 1.000000001 \ ]$$

In the following example we compute the exponential of a matrix. This can be used in mathematics to solve a system of linear differential equations.

---

[4]Facilities for computing both symbolic and numeric eigenvectors are available.

```
> B := matrix(2,2,[22,28,-15,-19]);
                              [  22    28 ]
                         B := [           ]
                              [ -15   -19 ]

> exponential(B,t);
          [ - 20 exp(t) + 21 exp(2 t)  - 28 exp(t) + 28 exp(2 t) ]
          [                                                       ]
          [   15 exp(t) - 15 exp(2 t)    21 exp(t) - 20 exp(2 t)  ]
```

Maple knows about many kinds of special matrices, such as Vandermonde and Hilbert matrices.

```
> vandermonde([u,v,w]) ;
                              [          2 ]
                              [ 1   u   u  ]
                              [            ]
                              [          2 ]
                              [ 1   v   v  ]
                              [            ]
                              [          2 ]
                              [ 1   w   w  ]

> factor(det (")) ;
                    (- w + u) (- w + v) (v - u)

> H := hilbert(2,x) ;
                              [   1       1   ]
                              [ -----   ----- ]
                              [ 2 - x   3 - x ]
                         H := [               ]
                              [   1       1   ]
                              [ -----   ----- ]
                              [ 3 - x   4 - x ]

> inverse(H) ;
      [              2                                          ]
      [   - (- 3 + x)  (- 2 + x)    (- 3 + x) (- 2 + x) (- 4 + x) ]
      [                                                          ]
      [                                          2               ]
      [ (- 3 + x) (- 2 + x) (- 4 + x)   - (- 3 + x)  (- 4 + x)    ]
```

There are operations for computing the Hermite and Smith canonical forms of matrices.

```
> smith(",x) ;
           [ - 3 + x              0            ]
           [                                   ]
           [                      2            ]
           [    0      (- 2 + x) (x  - 7 x + 12) ]
```

1.6 Equation Solving

1.6.1 General Equations

Maple has a "`solve`" function which can solve many kinds of equations, including single equations involving elementary transcendental functions, and systems of linear or polynomial equations.

Maple attempts to find all solutions for a polynomial equation.

```
> poly := 2*x^5 - 3*x^4 + 38*x^3 - 57*x^2 - 300*x + 450 ;
                  5     4       3        2
         poly := 2 x  - 3 x  + 38 x  - 57 x  - 300 x + 450
```

```
> solve(poly = 0, x) ;
                    1/2      1/2
             3/2, 6    , - 6    , 5 I, - 5 I
```

```
> solve(a*x^2/2 + b*x + c = 0, x) ;
            2         1/2         2        1/2
       - b + (b  - 2 a c)    - b - (b  - 2 a c)
       ---------------------, ---------------------
                 a                      a
```

Here we solve a system of four linear equations in four unknowns.

```
> e1 := 3*r + 4*s - 2*t + u = -2:
> e2 := r - s + 2*t + 2*u = 7:
> e3 := 4*r - 3*s + 4*t - 3*u = 2:
> e4 := -r + s + 6*t - u = 1:
> SolutionSet := solve({e1, e2, e3, e4}, {r, s, t, u}) ;
          SolutionSet := {t = 3/4, s = -1, r = 1/2, u = 2}
```

Using the "`subs`" command to simultaneously substitute the solutions for r, s, t, u back into the original equations, we can check the validity of solve.

```
> subs(SolutionSet, {e1, e2, e3, e4}) ;
                {2 = 2, 1 = 1, -2 = -2, 7 = 7}
```

Since exact arithmetic is used, Maple can properly handle non-square systems. And in the case where a system is under-determined, Maple will parameterize the solution space by treating one or more of the unknowns as free variables..

```
> solve( {e1, e2, e3-e1}, {r, s, t, u} );
          18       123                                29
   {t = ---- s + ---, r = - 9/17 s - 1/34, u = - 5/17 s + ----, s = s}
          17       68                                17
```

```
> subs( ", {e1, e2, e3-e1} );
               {-2 = -2, 7 = 7, 4 = 4}
```

Equations can have coefficients which are parameters.

```
> eqns := {x + y + z = a, x + 2*y - a*z = 0, b*z + a*y = 0} :
> solve(eqns, {x, y, z}) ;
```

$$\{z = \frac{a^2}{a + a^2 + b}, \ y = - \frac{a \ b}{a + a^2 + b}, \ x = \frac{a \ (2 \ b + a^2)}{a + a^2 + b}\}$$

And Maple includes algorithms for solving systems of polynomial equations.

```
> eqns := {x^2 + y^2 = 1, x^2 + x = y^2} :
> solve(eqns, {x, y}) ;
```

$$\{y = 0, x = -1\}, \ \{y = 0, x = -1\}, \ \{x = 1/2, y = 1/2 \ 3^{1/2}\},$$

$$\{y = - 1/2 \ 3^{1/2}, x = 1/2\}$$

1.6.2 Differential Equations

Whereas **solve** is used to solve algebraic equations, the **dsolve** function is used to solve differential equations. Maple can solve many first and second order differential equations.

```
> de1 := x^2 * diff(y(x), x) + y(x) = exp(x) ;
```

$$de1 := x^2 \left(\frac{d}{dx} y(x)\right) + y(x) = exp(x)$$

```
> dsolve(de1, y(x)) ;
```

$$y(x) = exp(1/x) \left(\int \frac{exp\left(-\frac{(x - 1) \ (x + 1)}{x}\right)}{x^2} \ dx + exp(1/x) \ \_C1 \right)$$

Notice that in the result above, Maple has denoted the arbitrary parameter in the solution by _C1. We can specify initial conditions. In our example below we have also simplified the result.

```
> de := diff(y(t), t) + y(t)^2 + (2*t + 1) * y(t) + t^2 + t + 1 = 0 ;
```

$$de := \left(\frac{d}{dt} y(t)\right) + y(t)^2 + (2 \ t + 1) \ y(t) + t^2 + t + 1 = 0$$

```
> ic := y(1) = 1 ;
```

$$ic := y(1) = 1$$

```
> simplify(dsolve({de, ic}, y(t))) ;
y(t) =
```

$$- \frac{- 3 \ t \ exp(- 1 - 1/2 \ t^2) + 2 \ exp(- 1/2 \ t \ (t + 2)) \ t + 2 \ exp(- 1/2 \ t \ (t + 2))}{- 3 \ exp(- 1 - 1/2 \ t^2) + 2 \ exp(- 1/2 \ t \ (t + 2))}$$

Maple can also find series approximations to solutions of differential equations.

```
> deqn := x^2 * diff(y(x),x$2) + x*diff(y(x),x) + (x^2 - u^2)*y(x) = 0;
                / 2     \
             2 |  d      |     / d      \     2    2
    deqn := x  |----- y(x)| + x |---- y(x)| + (x  - u ) y(x) = 0
             |   2     |     \ dx     /
              \ dx     /
```

```
> dsolve(deqn,y(x),series);
              (- u) /      1      2            1           4      6 \
    y(x) = _C1 x      |1 + ------- x  + -------------------- x  + O(x )|
                  \    4 u - 4     (8 u - 16) (4 u - 4)          /

          u /      1       2             1            4      6 \
    + _C2 x  |1 + --------- x  + ------------------------ x  + O(x )|
           \    - 4 u - 4     (- 8 u - 16) (- 4 u - 4)          /
```

1.6.3   Recurrence Equations

Here are some example solutions to recurrence equations using the command "**rsolve**".

```
> rsolve(s(n) = -3*s(n-1) - 2*s(n-2), s(n)) ;
                             n              n
              (2 s(0) + s(1)) (-1)  + (- s(0) - s(1)) (-2)
```

```
> rsolve({s(n) = -3*s(n-1) - 2*s(n-2), s(1) = 1, s(2) = 1}, s(n)) ;
                          n       n
                   - 3 (-1)  + (-2)
```

```
> rsolve(s(n) = 3*s(n/2) + 5*n, s(n)) ;
                   /ln(3)\ /          /ln(n)    \     \
                   |-----| |          |----- + 1|     |
                   \ln(2)/ |          \ln(2)    /     |
              n         \- 15 (2/3)              + 10/
```

Even when Maple cannot find an explicit solution, one can often apply other Maple tools.

```
> rsolve( x(n+1)=sin(x(n)), x(n) );
                rsolve(x(n + 1) = sin(x(n)), x(n))
```

```
> asympt(", n);
                              1/2    1
                      _C + 3/5 3    ln(----)
               1/2                    1/2
               3                      n            1
             ---- + -------------------- + O(----)
               1/2            3/2              2
               n              n                n
```

```
> rsolve( x(n+1)=(n-1)*x(n)-x(n-1)/(n+1), x(n) );
                             x(n - 1)
              rsolve(x(n + 1) = (n - 1) x(n) - --------, x(n))
                             n + 1
```

```
> asympt(", n,4);
                           /     1      7       1   \
               _C1 GAMMA(n - 1) |1 + ---- + ---- + O(----)|
                           |       2      3       4   |
                           \     2 n    6 n      n   /
```

1.6.4   Other Solvers

Solve attempts to find all solutions symbolically. Sometimes you may wish to find only numerical solutions, perhaps in a specific range. The "fsolve" function will generally return only one numerical result, although for polynomials it finds all real roots numerically (or all complex roots if the option 'complex' is specified).

```
> fsolve(x^4 - 2, x) ;
                            -1.189207115, 1.189207115

> fsolve(x^4 - 2, x, complex) ;
                -1.189207115, 1.189207115 I,  - 1.189207115 I, 1.189207115

> fsolve(x = cos(x), x) ;
                                   .7390851332

> eq1 := sin(x+y) - exp(x)*y = 0:
> eq2 := x^2-y = 2:
> fsolve({eq1,eq2},{x,y},{x=-1..1,y=-2..0});
                     {y = -1.552838698, x = -.6687012050}
```

Maple can also solve equations in **Z** mod p with the msolve function and equations over **Z** with the isolve function.

1.7   Output and Programming

The linear algebra package provides a function to compute the Jacobian of a vector function.

```
> f := 1+cos(x-y)/sin(x+y):
> g := 1-cos(x+y)/sin(x-y):
> J := linalg[jacobian]([f,g],[x,y]) ;

        [   sin(x - y)    cos(x - y) cos(x + y)   sin(x - y)    cos(x - y) cos(x + y) ]
        [ - ---------- - -------------------- -   ---------- - -------------------- ]
        [   sin(x + y)               2           sin(x + y)               2         ]
        [                     sin(x + y)                            sin(x + y)       ]
  J := [                                                                             ]
        [   sin(x + y)    cos(x + y) cos(x - y)   sin(x + y)    cos(x + y) cos(x - y) ]
        [   ---------- + --------------------     ---------- - -------------------- ]
        [   sin(x - y)               2           sin(x - y)               2         ]
        [                     sin(x - y)                            sin(x - y)       ]
```

1.7.1   LaTeX Output

Maple has several output forms which allow you to include Maple output in documents. For example, to produce LaTeX code for the Jacobian matrix above, one does

```
> latex(");
\left [\begin {array}{cc} -{\frac {\sin(x-y)}{\sin(x+y)}}-{\frac {\cos
(x-y)\cos(x+y)}{\sin(x+y)^{2}}}&{\frac {\sin(x-y)}{\sin(x+y)}}-{\frac
{\cos(x-y)\cos(x+y)}{\sin(x+y)^{2}}}\\\noalign{\medskip}{\frac {\sin(x
+y)}{\sin(x-y)}}+{\frac {\cos(x-y)\cos(x+y)}{\sin(x-y)^{2}}}&{\frac {
\sin(x+y)}{\sin(x-y)}}-{\frac {\cos(x-y)\cos(x+y)}{\sin(x-y)^{2}}}
\end {array}\right ]
```

This output when included in this document produces

$$
\left[
\begin{array}{cc}
-\dfrac{\sin(x-y)}{\sin(x+y)} - \dfrac{\cos(x-y)\cos(x+y)}{\sin(x+y)^2} & \dfrac{\sin(x-y)}{\sin(x+y)} - \dfrac{\cos(x-y)\cos(x+y)}{\sin(x+y)^2} \\[2ex]
\dfrac{\sin(x+y)}{\sin(x-y)} + \dfrac{\cos(x-y)\cos(x+y)}{\sin(x-y)^2} & \dfrac{\sin(x+y)}{\sin(x-y)} - \dfrac{\cos(x-y)\cos(x+y)}{\sin(x-y)^2}
\end{array}
\right]
$$

1.7.2   Fortran and C Code

Maple can also generate Fortran and C code. Here we generate "optimized" Fortran code.

```
> fortran(J,optimized);
      t2 = x-y
      t3 = sin(t2)
      t4 = x+y
      t5 = sin(t4)
      t7 = t3/t5
      t9 = cos(t2)
      t10 = t5**2
      t12 = cos(t4)
      t15 = -t9/t10*t12
      t19 = t5/t3
      t20 = t3**2
      t23 = t12/t20*t9
      J(1,1) = -t7+t15
      J(1,2) = t7+t15
      J(2,2) = t19-t23
      J(2,1) = t19+t23
```

How good is the optimizer? We can compare costs[5] by

```
> readlib(cost): readlib(optimize):
> cost(J);
 24 additions + 16 multiplications + 20 functions + 8 divisions + 4 assignments

        + 4 subscripts
```

---

[5]The **readlib** command loads these specialized procedures into our session.

```
> cost(optimize(J));
   6 additions + 16 assignments + 4 functions + 4 divisions + 9 multiplications

      + 4 subscripts
```

1.7.3 Programming In Maple

An important component of the Maple system is the Maple programming language which may be
used to write procedures. Following are some examples of procedures written in Maple.

A procedure to compute the Fibonacci numbers. The purpose of "option remember" is to tell the
system to store computed values as it proceeds.

```
> F := proc(n) option remember; F(n) := F(n-1) + F(n-2) end:
> F(0) := 0:
> F(1) := 1:
```

Some examples invoking procedure F:

```
> F(101) ;
```
$$573147844013817084101$$

```
> seq( F(i), i=1..10 ) ;
                        1, 1, 2, 3, 5, 8, 13, 21, 34, 55
```

A routine to compute the standard deviation of a list of numbers.

```
> sigma := proc(data) local mean,n,s,x;
>       n := nops(data);
>       if n < 2 then ERROR(`input must contain at least 2 values`) fi;
>       mean := 0;
>       for x in data do mean := mean + x od;
>       mean := mean/n;
>       s := 0;
>       for x in data do s := s + (x-mean)^2 od;
>       sqrt(s/(n-1))
> end:
```

Some examples illustrating exact and numerical computation.

```
> sigma( [1,2,3,4,5,6] );
```
$$\frac{1}{2} \, 7^{1/2} \, 2^{1/2}$$

```
> sigma( [0.5,3.2,5.1] );
```
$$2.311565127$$

Compute the Chebyshev polynomials of the first kind, $T_n(x)$ of degrees 0 through n, and store them
in a table. This time we've included some argument checking to make our program more robust.

```
> ChebyshevTable := proc(n) local p,k ;
>       if not type(n,integer) or n < 1 then
>                ERROR(`1st argument must be a positive integer`)
>       fi;
>       p[0] := 1;  p[1] := x;
>       for k from 2 to n do
```

```
>                       p[k] := expand( 2*x*p[k-1] - p[k-2] )
>              od;
>              RETURN( eval(p) )
> end:
```

An example invoking procedure `ChebyshevTable`:

```
> a := ChebyshevTable(5):
> a[0], a[1], a[2], a[3], a[4], a[5] ;
```

$$1, \; x, \; 2\,x^2 - 1, \; 4\,x^3 - 3\,x, \; 8\,x^4 - 8\,x^2 + 1, \; 16\,x^5 - 20\,x^3 + 5\,x$$

The Chebyshev polynomials are orthogonal on the interval $[-1, 1]$ in the sense that given two Chebyshev polynomials $T_i(x)$ and $T_j(x)$ where $i <> j$,

$$\int_{-1}^{1} \frac{T_i(x) T_j(x)}{\sqrt{1 - x^2}} \, dx = 0$$

Here is a Maple program to test this statement. Or perhaps we should say, a program to test Maple's integration command!

```
> check := proc(T,n) local i,j,r;
>         for i to n do
>             for j to i-1 do
>                 r := int( T[i]*T[j]/sqrt(1-x^2), x=-1..1 );
>                 if r <> 0 then RETURN( (i,j) = r ) fi
>             od
>         od;
>         print(`all okay!`)
> end:
> check(a,5);
```

$$\text{all okay!}$$

Finally, here is an example of a procedure defined in the Maple library which computes the least common multiple of a sequence of polynomials. The command "`interface`" with `verboseproc=2` allows the user to print code for any library function.

```
> interface(verboseproc=2) ;
> print(lcm) ;
proc(a,b)
local q;
options remember,`Copyright 1990 by the University of Waterloo`;
    if nargs = 0 then 1
    elif nargs = 1 then expand(a); sign(")*"
    elif 2 < nargs then lcm(a,lcm(op(2 .. nargs,[args])))
    elif type(a,integer) and type(b,integer) then ilcm(a,b)
    else gcd(a,b,q); expand(q*b); sign(")*"
    fi
end
```

Unlike the previous procedures, `lcm` may be invoked with an arbitrary number of arguments. The sequence of actual arguments is the value of the special name **args**. The number of actual arguments is the value of the special name **nargs**.

"quit", "done" or "stop" will terminate a Maple session.

```
> quit
bytes used=21260952, alloc=1712128, time=31.066
```

The actual number of bytes used and time used will vary from system to system.

2
Language Elements

2.1 Character Set

The Maple character set consists of letters, digits, and special characters. The letters are the 26 *lower case letters*

    a,b,c,d,e,f,g,h,i,j,k,l,m,n,o,p,q,r,s,t,u,v,w,x,y,z

and the 26 *upper case letters*

    A,B,C,D,E,F,G,H,I,J,K,L,M,N,O,P,Q,R,S,T,U,V,W,X,Y,Z .

The 10 *digits* are

    0,1,2,3,4,5,6,7,8,9

and there are 32 *special characters*, as shown in Table 2.1. There are two-character alternates for five of these special characters, for use when the complete ASCII character set is not available; these are shown in Table 2.2.

2.2 Tokens

The tokens consist of keywords, programming-language operators, strings, natural integers, and punctuation marks. The *keywords* in Maple are listed in Table 2.3.

There are three types of *programming-language operators*, namely the *binary*, *unary*, and *nullary* operators. These are listed in Tables 2.4, 2.5, and 2.6, respectively. Note that seven of these operators are reserved words: mod, union, minus, intersect, and, or, not .

The simplest instance of a *string* is a letter followed by zero or more letters, digits, and underscores. Another instance of a string is the underscore followed by zero or more letters, digits, and underscores. More generally, a string can be formed by enclosing any sequence of characters in back quotes (grave accents). In all cases, the maximum length of a string in Maple is 499 characters. A string is a valid name (such as a variable name or a function name) but the user should avoid using names which begin with an underscore since such names are created and used globally by the Maple system. A Maple string is also used in the sense of a "character string", usually by enclosing a sequence of characters in back quotes. (See Section 3.2.2 for more details and examples on string and name formation, including the issue of enclosing special characters such as

	blank	(	left parenthesis
;	semicolon	)	right parenthesis
:	colon	[	left bracket
+	plus	]	right bracket
−	minus	{	left brace
*	asterisk	}	right brace
/	slash	`	grave accent (back quote)
^	caret (circumflex)	'	apostrophe (single quote)
!	exclamation	"	double quote (ditto)
=	equal	|	vertical bar
<	less than	&	ampersand
>	greater than	_	underscore
@	at sign	%	percent
$	dollar	\	backslash
.	period	#	sharp
,	comma	?	question mark

TABLE 2.1. Special characters

Character		Alternate
^	caret (circumflex)	**
[	left bracket	(|
]	right bracket	|)
{	left brace	(*
}	right brace	*)

TABLE 2.2. Character alternates

by	do	done	elif
else	end	fi	for
from	if	in	local
od	option	options	proc
quit	read	save	stop
then	to	while	

TABLE 2.3. Maple keywords

Operator	Meaning	Operator	Meaning
+	addition	<	less than
−	subtraction	<=	less or equal
*	multiplication	>	greater than
/	division	>=	greater or equal
**	exponentiation	=	equal
^	exponentiation	<>	not equal
$	sequence operator	->	arrow operator
@	composition	mod	modulo
@@	repeated composition	union	set union
&*	noncommutative mult.	minus	set difference
&string	neutral operator	intersect	set intersection
.	concatenation; decimal	and	logical and
..	ellipsis	or	logical or
,	expression separator	:=	assignment

TABLE 2.4. Binary operators

Operator	Meaning
+	unary plus (prefix)
−	unary minus (prefix)
!	factorial (postfix)
$	sequence operator (prefix)
not	logical not (prefix)
&string	neutral operator (prefix)
.	decimal point (prefix or postfix)
%	label (prefix)

TABLE 2.5. Unary operators

Operator	Meaning
"	latest expression
" "	penultimate expression
" " "	before penultimate expression

TABLE 2.6. Nullary operators

;	semicolon	(	left parenthesis
:	colon	)	right parenthesis
'	single quote	[	left bracket
`	back quote	]	right bracket
|	vertical bar	{	left brace
<	left angle bracket	}	right brace
>	right angle bracket		

TABLE 2.7. Maple punctuation marks

back quote, backslash, or newline within a string.)

A *natural integer* is any sequence of one or more digits. The numeric constants in Maple (integers, rational numbers, and floating-point numbers) are formed from the natural integers using programming-language operators. The length of a natural integer, and hence the length of integers, rational numbers, and floating-point numbers, is arbitrary (that is, the length limit is system-dependent but generally much larger than users will encounter).

The *punctuation marks* are listed in Table 2.7. The semicolon and the colon are used to separate statements. The distinction between these marks is that during an interactive session if a statement is followed by a colon, the result of the statement is not printed. Enclosing an expression in a pair of single quotes specifies that the expression is to be left unevaluated. The back quote is used to form strings. The left and right parentheses are used to group terms in an expression and to group parameters in a function call. The left and right brackets are used to form indexed (subscripted) names and to select components from aggregate objects such as arrays, tables, and lists. The left and right brackets are also used to form lists and the left and right braces are used to form sets. The left and right angle brackets are used to form functional operators, as is the vertical bar which serves to separate the body of the functional operator from the formal parameters and local variables (see Chapter 8).

2.3   Escape Characters

The *escape characters* are ?, !, #, and \ .

The ? character must appear as the first non-blank character on a line, in which case it is translated into an invocation of Maple's *help* facility. The words following ? on the same line determine the arguments to the help procedure.

The ! character, if it appears as the first non-blank character on a line, is treated as an *escape-to-host* operator. The remainder of the line is passed as a command to the host operating system (see Chapter 11).

The # character is used to indicate that the following characters on the line are to be treated as a *comment*. The \ character is used for *continuation* of lines and for grouping of characters within a token. See the following section for more details about the latter two escape characters.

2.4   Blanks, Lines, Comments, and Continuation

The *white space* characters are ⟨space⟩, ⟨tab⟩, ⟨return⟩, and ⟨line feed⟩. We will use the terminology *newline* to refer to either ⟨return⟩ or ⟨line feed⟩ since these characters are not distinguished in the Maple system. Whenever the terminology *blank* is used, it is understood to mean either ⟨space⟩ or ⟨tab⟩. The white space characters separate tokens, but are not themselves tokens. White space characters cannot occur within a token but otherwise may be used freely. The one exception to the free use of white space characters is the formation of a string by enclosing a sequence of characters within back quotes, in which case the white space characters are significant like any other character.

Input to the Maple system consists of a sequence of statements separated by the statement separators, which are the semicolon and the colon characters. The system operates in an interactive mode, executing statements as they are entered. A *line* consists of a sequence of characters followed by a newline character. A single line may contain several statements, an incomplete statement (to be completed on succeeding lines), or several statements followed by an incomplete statement. When a line is entered, the system evaluates (executes) the statements which have been completed on that line (if any). At the interactive level, a statement will be recognized as complete only when a statement separator is encountered. (One exception is the `quit` statement which need not be followed by a statement separator; also, when using the escape characters ? or ! it is not necessary to end the command with a statement separator.)

On a line, all characters which follow a sharp character (#) are considered to be part of a *comment*. However, the sharp character is not treated as the beginning of a comment if it is enclosed within a pair of back quotes (in which case it is a character within a string).

Since a newline character may be used freely anywhere except within a token, statements may be continued from line to line quite naturally. The problem of continuation from one line to the next is less trivial when long numbers or long strings are involved, since these are the two classes of tokens which are not restricted to being only a few characters in length. The general mechanism in Maple to specify continuation of one line onto the next line is as follows: if a newline character is immediately preceded by the special character backslash (\ ), then both the backslash and the newline are ignored by the parser. Another property of the backslash character is that it is itself ignored if it is immediately followed by any character other than newline. This is useful when it is desired to break up a long sequence of digits into groups of smaller sequences to enhance readability. (See Section 3.2.1 for examples of specifying long numbers using grouping of digits, and using continuation onto a new line.)

For the case of strings formed as a sequence of characters enclosed within back quotes, the blank and newline characters are all valid characters within a string (although, as a safety feature, a warning message is generated whenever a newline is encountered within a string). If it is desired to make the newline transparent in the specification of a string (by allowing continuation of the string specification onto a new line rather than the insertion of a newline character into the string) then the general mechanism discussed above must be used: the newline must be immediately preceded by the special character backslash (\ ). The backslash character is itself normally ignored, as noted above, but if two backslash characters occur in succession then only the first one is ignored, yielding a mechanism for specifying backslash as one of the characters in a string. (See Section 3.2.2 for examples of string specification.)

2.5   Files

The file system is an important part of the Maple system. The user interacts with the file system either explicitly by way of the **read** and **save** statements, or implicitly by specifying a function name corresponding to a file which the system will read in automatically. The file naming conventions used within the Maple system do not necessarily correspond directly with the file naming conventions on the host operating system. A translation from Maple's filenames to the host system's filenames takes place automatically. The **convert** function may be used in the form

> `convert( ⟨Maple filename⟩, hostfile )`

to cause Maple to display the actual filename that will be used on the host system.

A *file* consists of a sequence of statements either in "Maple internal format" or in "user format." If the file is in user format then the effect of reading the file is identical to the effect of the user entering the same sequence of statements. The system will display the result of executing each statement which is read in from the file. (The user can control the echoing of the input lines being read via the setting of the **echo** variable of the **interface** command — see **?interface**.) On the other hand, if the file is in Maple internal format then reading the file causes no information to be displayed to the user but updates the current Maple environment with the contents of the file. Maple assumes that a file will be in Maple internal format when its filename ends with the characters '.m'. For example, some typical names for files in user format are:

```
temp
`lib/src/gcd`
```

while some typical names for files in internal format are:

```
`temp.m`
`lib/gcd.m`
```

(Note that filenames involving characters such as '/' or '.' must be enclosed in back quotes in order to be interpreted properly as ⟨name⟩'s.)

There are two ways to create a file in user format. One way is to use a text editor on the host system to create a file in which Maple statements have been entered just as they would be entered in an interactive Maple session. Another way is to use the Maple **save** statement from within a Maple session, saving into a file whose name does not have the '.m' suffix. The latter action writes the values of specified variables (or all assigned variables if none are specified) into the file as user-readable Maple assignment statements.   A file in Maple internal format can be created only by using the **save** statement from within a Maple session, saving into a file whose name has the suffix '.m'. Internal-format files are not in a form that can be read normally; they are meant to be read only by the Maple system. Either type of file may be read (loaded) into a Maple session by using the Maple **read** statement.   Many Maple functions are not part of the basic Maple system which is loaded initially, but rather reside in files in the Maple library. When one of these functions is invoked in Maple, the internal-format file containing the function definition is automatically loaded. (See the *Maple V Library Reference Manual*.)

3
Statements and Expressions

3.1 Types of Statements

There are eight types of statements in Maple. They will be described informally here. The formal syntax is given in Section 3.3.

3.1.1 Assignment Statement

The form of this statement is

$$\langle \text{name} \rangle \ := \ \langle \text{expression} \rangle$$

and it associates a name with the value of an expression.

3.1.2 Expression

An ⟨expression⟩ is itself a valid statement. The effect of this statement is that the expression is evaluated.

3.1.3 Read Statement

The statement

read ⟨expression⟩

causes a file to be read into the Maple session. The ⟨expression⟩ must evaluate to a name which specifies the file. The file name may be one of two types as discussed in Section 2.5. A typical example of a **read** statement is

```
read `lib/f.m`
```

where the back quotes are necessary so that the expression evaluates to a name.

If the file is in Maple internal format[1], the file is simply read. Any Maple data or procedures in the file are now accessible. If the file is in user format, it must contain syntactically valid Maple statements. The file is read and each statement in the file is executed as if it had been typed into the Maple session by the user. The user can control the echoing of the input lines being read via the setting of the `echo` variable of the `interface` command (see `?interface`).

3.1.4   Save Statement

The **save** statement has a form corresponding to the **read** statement

> **save** ⟨expression⟩

and it can also be used in the more general form

> **save** ⟨nameseq⟩ , ⟨expression⟩ .

The simpler form causes the current Maple environment to be written into a file. The ⟨expression⟩ must evaluate to a name which specifies the file. If the file name ends with the characters '.m' then the environment is saved in Maple internal format; otherwise the environment is saved in user format as a sequence of assignment statements.

The generalized syntax for the save statement allows the specification of a sequence of specific names to be saved, yielding a *selective save* facility. It takes the general form

> **save** ⟨name⟩$_1$, ⟨name⟩$_2$, ... , ⟨name⟩$_k$, ⟨filename⟩

where ⟨name⟩$_1$, ... , ⟨name⟩$_k$ are the names of assigned variables to be saved and ⟨filename⟩ evaluates to the name of a file. Each argument except the last one is "evaluated to a name" while the last argument is fully evaluated. The construct ⟨name⟩.(⟨range⟩) may be used here to construct a sequence of names. (See Section 3.2.9.)

3.1.5   Selection Statement

The selection statement takes one of the following general forms. Here ⟨expr⟩ is an abbreviation for ⟨expression⟩ and ⟨statseq⟩ stands for a sequence of statements.

> **if** ⟨expr⟩ **then** ⟨statseq⟩ **fi**
> **if** ⟨expr⟩ **then** ⟨statseq⟩ **else** ⟨statseq⟩ **fi**
> **if** ⟨expr⟩ **then** ⟨statseq⟩ **elif** ⟨expr⟩ **then** ⟨statseq⟩ **fi**
> **if** ⟨expr⟩ **then** ⟨statseq⟩ **elif** ⟨expr⟩ **then** ⟨statseq⟩ **else** ⟨statseq⟩ **fi**

---

[1] A file is in Maple internal format if the name of the file ends with `.m` .

Wherever the construct 'elif ⟨expr⟩ **then** ⟨statseq⟩' appears in the above forms, this construct may be repeated any number of times to yield a valid selection statement. The sequence of statements in the branch selected (if any) is executed. The construct 'elif ⟨expr⟩ **then** ⟨statseq⟩' has the meaning '**else if** ⟨expr⟩ **then** ⟨statseq⟩', but the latter form would require a closing **fi** delimiter for each opening **if**. Use of the **elif** keyword avoids the need for multiple **fi** delimiters. The selection statement can be viewed as a form of a "case statement" since multiple conditions can be specified, and clearly the **else** clause is the "default case".

Each ⟨expr⟩ appearing in an **if**-part or an **elif**-part of a repetition statement will be evaluated as a Boolean expression. The evaluation of Boolean expressions in Maple uses *three-valued logic* (see Section 3.2.8). In addition to the special names **true** and **false**, Maple also understands the special name **FAIL**. The value **FAIL** is sometimes used as the value returned by a procedure when it is unable to completely solve a problem. In other words, it can be viewed as the value "don't know". The truth tables for three-valued logic are presented in Section 3.2.8.

3.1.6   Repetition Statement

The syntax of the repetition statement is either of the following two forms (⟨expr⟩ and ⟨statseq⟩ are defined in the previous section):

> **for** ⟨name⟩ **from** ⟨expr⟩ **by** ⟨expr⟩ **to** ⟨expr⟩ **while** ⟨expr⟩ **do** ⟨statseq⟩ **od**
> **for** ⟨name⟩ **in** ⟨expr⟩ **while** ⟨expr⟩ **do** ⟨statseq⟩ **od**

where any of '**for** ⟨name⟩', '**from** ⟨expr⟩', '**by** ⟨expr⟩', '**to** ⟨expr⟩', or '**while** ⟨expr⟩' may be omitted. The sequence of statements in ⟨statseq⟩ is executed zero or more times. The '**for** ⟨name⟩' part may be omitted if the index of iteration is not required in the loop, in which case a "dummy index" is used by the system.

In the first form above, if the '**from** ⟨expr⟩' part or the '**by** ⟨expr⟩' part is omitted then the default value '**from** 1' or '**by** 1', respectively, is used. If the '**to** ⟨expr⟩' part or the '**while** ⟨expr⟩' part is present then the corresponding tests for termination are checked at the beginning of each iteration, and if neither is present then the loop will be an infinite loop unless it is terminated by the **break** construct (see below), or by a **return** from a procedure (see Section 7.7), or by execution of the **quit** statement (see Section 3.1.8).

In the second form of the repetition statement, the index (the ⟨name⟩ specified in the for-part of the statement) takes on, in order, values from the list [**op**(⟨expr⟩)] where ⟨expr⟩ is the expression specified in the '**in** ⟨expr⟩' part of the statement (or if ⟨expr⟩ is itself an *expression sequence* then the index simply takes on values from that sequence). In other words, the successive values of the index are the successive operands of ⟨expr⟩. (See Section 4.1 for a description of the operands associated with each Maple data type.) The value of the index may be tested in the while-part of the statement and, of course, the value of the index is available when executing the ⟨statseq⟩. As in the first form, if the '**while** ⟨expr⟩' part is present then the test for termination is checked at the beginning of each iteration.

The ⟨expr⟩ appearing in the **while**-part of a repetition statement will be evaluated as a Boolean expression. As mentioned in the preceding subsection, the evaluation of boolean expressions in Maple uses *three-valued logic* (see Section 3.2.8).

There are two additional loop control constructs in the Maple language: **break** and **next**. When the special name **break** is evaluated, the result is to exit from the innermost repetition statement within which it occurs. Execution then proceeds with the first statement following this repetition statement. When the special name **next** is evaluated, the result is to exit from the current ⟨statseq⟩ corresponding to the innermost repetition statement within which it occurs, and to cause execution to proceed with the next iteration of this repetition statement. Note that "to proceed with the next iteration" implies updating the index of iteration and then applying the tests for termination (if any) before proceeding, so an exit from the repetition statement will occur if the tests for termination so indicate. It is an error if the names **break** or **next** are evaluated in a context other than within a repetition statement.

3.1.7 Empty Statement

The empty statement is syntactically valid in Maple. For example,

```
a := 1; ; quit
```

is a valid statement sequence in Maple consisting of an assignment statement, the empty statement, and the **quit** statement. Of course since blanks may be used freely, any number (including zero) of blanks could appear between the semicolons, yielding a statement sequence syntactically identical to the one presented here.

3.1.8 Quit Statement

The syntax of the **quit** statement is any one of the following three forms:

>   **quit**
>   **done**
>   **stop**

The result of this statement is to terminate the Maple session and return the user to the system level from which Maple was entered. (In many implementations of Maple, pressing the break/interrupt key twice in rapid succession, or some similar system-specific sequence, can also be used to cause termination of a Maple session.)

3.2 Expressions

Expressions are the fundamental entities in the Maple language. The various types of expressions are described informally here. The formal syntax is given in Section 3.3.

3.2.1 Constants

The numeric constants in Maple are integers, fractions, and floating-point numbers. A ⟨natural integer⟩ is any sequence of one or more digits of arbitrary length (the length limit is system-dependent but generally much larger than users will encounter). An *integer* is a ⟨natural integer⟩ or a signed integer (indicated by +⟨natural integer⟩ or -⟨natural integer⟩). A *fraction* is of the form

⟨integer⟩ / ⟨natural integer⟩ .

(Note that a fraction is always simplified by Maple so that the denominator is positive, and it will also be reduced to lowest terms).

An ⟨unsigned float⟩ is one of the following three forms:

⟨natural integer⟩ . ⟨natural integer⟩
⟨natural integer⟩ .
. ⟨natural integer⟩

A *floating-point number* is an ⟨unsigned float⟩ or a signed float (indicated by +⟨unsigned float⟩ or -⟨unsigned float⟩). The **evalf** function is used to force an expression to be evaluated to a floating-point number (if possible). The number of digits carried in the 'mantissa' when evaluating floating-point numbers is determined by the value of the global name **Digits** which has 10 as its default value.

It is sometimes necessary to specify a numeric constant whose length exceeds the length of a single line. Since a newline character is not allowed within a ⟨natural integer⟩ token, Maple's general continuation mechanism (see Section 2.4) must be used when line-breaking within a sequence of digits. Namely, the newline character must be immediately preceded by the special character backslash (\ ), in which case both the backslash and the newline character will be ignored. Some examples are given below. Since another property of the backslash character is that it is itself ignored if it is immediately followed by any character other than newline, the following examples also include cases where we have chosen to group the digits into groups of three to enhance readability. For example, the following two forms of input yield the same floating-point number:

```
3.141592653589793
3.141\592\653\589\793
```

and two additional forms of input yielding this same number are:

```
3.
141592653589793
```

```
3.141592653\
589793
```

In the four forms of input above, the first form uses no line-breaking and no grouping of digits. The second form uses the transparent backslash character to break the sequence of digits into groups of three, while the third form breaks the line at the decimal point, in which case it is not necessary to precede the newline character with a backslash character. The fourth form uses a line-break in the middle of a sequence of digits, in which case it is necessary to precede the newline with a backslash.

An alternative input format for floating-point numbers in Maple is to use the notation

```
Float(mantissa, exponent)
```

which directly corresponds to Maple's internal data structure representation of floating-point numbers. In this notation, both arguments to the **Float** function must be integers and the value is interpreted to be the floating-point number

$$\text{mantissa} \times 10^{\text{exponent}} .$$

For example, **Float(3141592654,-9)** represents the same floating-point number as the more standard input notation **3.141592654**, and it is also the value of any of the following input expressions:

```
3141592654. * 10^(-9)
.3141592654 * 10^1
314.1592654 * 10^(-2)
```

When Maple's line-printing mode is used (via the **lprint** function, or in normal display when pretty-printing is turned off via **interface(prettyprint=false)**)   the **Float(mantissa, exponent)** notation is used for floating-point numbers with very small or very large magnitudes. For example,

```
interface(prettyprint = false);  evalf(exp(-10));  evalf(exp(-20));
```

yields

```
Float(4539992976,-14)
Float(2061153622,-18)
```

in contrast with the normal display of these numbers which would be seen in pretty-printed format as:

```
.00004539992976

      -8
.2061153622*10
```

There is the concept of *symbolic constants* in Maple. The global variable **constants** is initially assigned the following expression sequence of names:

```
false, gamma, infinity, true, Catalan, E, Pi
```

implying that these particular names are understood by Maple to be of type *constant*. The user may define additional names (specifically, they must be the simplest type of names called *strings* — see Section 3.2.2) to be constants by redefining the value of this global variable. For example, if the assignment

```
constants := constants, g;
```
is executed, then the name **g** has been added to the constants sequence and `type(g,constant)` returns **true**. More generally, a Maple expression is of type *constant* if it is an unevaluated function with all arguments of type *constant*, or a sum, product, or power with all operands of type *constant*. For example, the following expressions are of type *constant* :

```
sin(1),  f(2,3),  exp(gamma),  4 + Pi,  2 * E / Pi^(1/2)  .
```

3.2.2   Names and Strings

A name in Maple has a value which may be any expression, or, if no value has been assigned to it, then it stands for itself. A common form of a name is a *string*, which in its simplest form is a letter followed by zero or more letters, digits, and underscores (with a maximum length of 499 characters). Note that there are 30 reserved words in the Maple system, consisting of 23 keywords plus 7 programming-language operators (see Section 2.2), which are not generally available as names formed in the above manner. Note also that lower case letters and upper case letters are distinct, so that the names

```
g    G    new_term    New_Term    x13a    x13A
```
are all distinct. Another instance of a *string* is the underscore followed by zero or more letters, digits, and underscores. Names beginning with the underscore are used as global variable names by the Maple system and therefore such names should be avoided by users.

A *string* can also be formed by enclosing any sequence of characters in back quotes. (To allow the back quote to be one of the characters in a string, two consecutive back quotes appearing after the opening of a string are parsed as an enclosed back quote character.) A reserved word enclosed in back quotes also becomes a valid Maple string, distinct from the usage of the reserved word as a token. The following are valid strings (and hence names) in Maple:

```
`This is a strange name:`    `2D`    `n-1`    `a``b`    `while`  .
```
The back quotes do not themselves form part of the string so they disappear when the string has been input to Maple. For example, if the fourth string in the above list is input to Maple, the response will be the string

```
`b .a
```
As another example, if **n** has the value 5 then the statement

```
`n-1` := n-1;
```
will yield the following response from Maple:

```
n-1 := 4  .
```
The user should beware of misusing this facility for string (name) formation to the point of writing unreadable programs!

The effect of a newline character within a string should be noted. A newline character is a valid character within a string. However, a warning message is issued whenever a newline occurs within a string. This warning message is a safety precaution to help the user identify the following type of bug. A common typographical mistake is to fail to close a string which has been opened, in which case all succeeding input lines are "eaten up" as part of the string. In this situation, the user finds that no matter what command is typed in the Maple session, there is no response from Maple except to prompt for more input, even when the **quit** statement is entered in an attempt to leave the Maple session out of frustration. With the warning message in place, every line of input yields the response:

```
Warning: String contains newline character. Close strings with ` quote.
```

For example, if the input to Maple is

```
interface(prettyprint = false);
s := `This is a
string`;
```

where the character following the word **a** is a newline, then the response from Maple is

```
Warning: String contains newline character. Close strings with ` quote.
s := This is a
string .
```

In this example, prettyprinting was turned off; with prettyprinting turned on, the line containing the assignment operator would be centered on the line. Note that the string assigned to **s** is 16 characters in length, one of the characters being the newline character.

Alternatively, it may be desired to specify a string not containing a newline character, but more than one input line is needed to specify the string. In this case, preceding the newline with a backslash will make the newline transparent, just as in the case of long numbers. For example, consider the statement:

```
if not type(n,integer) or n <= 0 then print(`The value of n is not a \
positive integer.`) fi;
```

If **n** has the value −2 when this statement is executed, then the result will be to print the string:

```
The value of n is not a positive integer.
```

In this case, the string does not contain the newline character. Finally, it should be noted that the backslash character can be entered as a character in a string by exploiting the fact that two successive backslash characters are parsed as a single backslash character (just as two successive back quotes are parsed as a single back quote). For example, the input

```
`Don't confuse the quotes:  ', ", and ``  , nor \
the slashes:  / and \ . ` ;
```

yields the following string as a response:

```
Don't confuse the quotes:  ', ", and `  , nor the slashes:  / and \ .
```

More generally, a ⟨name⟩ may be formed using the *concatenation operator* in one of the following three forms:

> ⟨name⟩ . ⟨natural integer⟩
> ⟨name⟩ . ⟨string⟩
> ⟨name⟩ . (⟨expression⟩)

Some examples of the use of the concatenation operator for ⟨name⟩ formation are:

```
v.5      p.n      a.(2*i)      V.(N.(i-1))      r.i.j
```

The concatenation operator is a binary operator which requires a ⟨name⟩ as its left operand. Its right operand is evaluated and then concatenated to the left operand. If the right operand is evaluated to an integer or a string, then the result of the concatenation is a name. If, on the other hand, the right operand is evaluated to some other type of object, say a floating-point number, then the result of the operation is an unevaluated concatenated object. (See *Unevaluated Concatenation* in Section 4.1.16.) For example, if n has the value 4 then p.n evaluates to the name p4, while if n has no value then p.n evaluates to the name pn. If n has the value 3.0 then p.n does not produce a name but instead yields an unevaluated concatenated object. If i has the value 5 then a.(2*i) evaluates to the name a10. As a final example, if N4 has the value 17 and i has the value 5 then V.(N.(i-1)) evaluates to the name V17, while V.N.(i-1) evaluates to the name VN4, assuming that N has no value. (See also the function cat in the Maple Library, for string concatenation.)

Yet another form of name in Maple is the *indexed name* which has the form

⟨name⟩ [⟨expression sequence⟩] .

Note that since an indexed name is itself a valid name, there can be a succession of subscripts as in:

```
A[1,2,3][x,y][2,1]  .
```

The use of the indexed name b[1] does not imply that b is an array, as is true in many other languages. The statement

```
a := b[1] + b[2] + b[1000];
```

simply forms the sum of three indexed names. It is not necessary that b have an array value. (However, if b does evaluate to an array or a table, then b[1] is the element of the array or table selected by 1.) The assignment of a value to an indexed expression will implicitly create a table. For example, consider the statement

```
b[1] := 10;
```

If b had no value previously, then this assignment statement creates a table for b and then assigns the value 10 to the component of this table indexed by 1. (See Chapter 6: Arrays and Tables.)

3.2.3 Labels

A *label* in Maple is formed by the unary operator % followed by a natural integer. A label is only valid after it has been introduced by Maple's pretty-printer for the output of an expression. The purpose is to allow the naming (labelling) of common subexpressions, which serves to decrease the size of the printed output and to make the output more comprehensible. After it has been introduced by the pretty-printer, a label may be used just like an assigned name in Maple.

Examples:

```
> f := exp(3*x^2-2*x+1)/(1+x^2):
> diff(f, x$2);
```

$$6\,\frac{\%1}{1+x^2} + \frac{(6x-2)^2\,\%1}{1+x^2} - 4\,\frac{(6x-2)\,\%1\,x}{(1+x^2)^2} + 8\,\frac{\%1\,x^2}{(1+x^2)^3} - 2\,\frac{\%1}{(1+x^2)^2}$$

$$\%1 := \exp(3x^2 - 2x + 1)$$

After the above printout has been obtained, the label %1 may be used like an assigned name and its value will be the expression `exp(3*x^2-2*x+1)`.

3.2.4 Expression Sequences

An *expression sequence* is an expression of the form

$$\langle expression \rangle_1 \, , \, \langle expression \rangle_2 \, , \, \ldots \, , \, \langle expression \rangle_n \, .$$

The comma operator is used to join expressions into an expression sequence. It has the lowest precedence of all operators except assignment (see Section 3.2.14). When expression sequences are formed, the result is simplified to a single, unnested expression sequence.

A zero-length expression sequence is syntactically valid. It arises, for example, in the context of forming an empty list, an empty set, a function call with no parameters, or an indexed name with no subscripts. The special name NULL is initially assigned the zero-length expression sequence and may be used when needed. Although the name NULL is an ordinary Maple ⟨name⟩, it should not be assigned a different value. Here's an example:

```
a := A, B, C;          ⟶      a := A, B, C
b := NULL;             ⟶      b :=
c := a, b, 1, 2, 3;    ⟶      c := A, B, C, 1, 2, 3
```

The function **seq** may be used to form expression sequences as follows. The syntax is either of the following general forms:

$$\textbf{seq}(\langle expr \rangle_1, \langle name \rangle = \langle expr \rangle_2 \,.. \, \langle expr \rangle_3)$$
$$\textbf{seq}(\langle expr \rangle_1, \langle name \rangle = \langle expr \rangle_2) \, .$$

In the first form, the result is the expression sequence produced by substituting for ⟨name⟩ in ⟨expr⟩$_1$ the values ⟨expr⟩$_2$, ⟨expr⟩$_2$ +1 , ... , ⟨expr⟩$_3$ (or up to the last value not exceeding ⟨expr⟩$_3$). If the value ⟨expr⟩$_2$ is greater than ⟨expr⟩$_3$ then the result is the NULL expression sequence.

If the second form is used, the result is the expression sequence produced by substituting for ⟨name⟩ in ⟨expr⟩$_1$ the values **op(1,**⟨expr⟩$_2$**)**, **op(2,** ⟨expr⟩$_2$**)**,... ,**op(nops(**⟨expr⟩$_2$**),** ⟨expr⟩$_2$**)**.

The `seq` function is related to the `for-loop` construct. It behaves in a manner similar to `for` except that `seq` constructs a sequence of values. The most typical usage is `seq(f(i), i = 1..n)` which generates the sequence `f(1)`, `f(2)`, ..., `f(n)`. More generally, `seq(f(i), i = m..n)` generates the sequence `f(m)`, `f(m+1)`, `f(m+2)`, ..., `f(n)`. Here m and n do not need to be integers. An invocation using the second form `seq(f(i), i = x)` generates a sequence by applying f to each operand of x. Here, x would most commonly be a `set` or `list`, but it could be any other data structure to which `op` can be applied, such as a `sum` or `product`.

The two versions of the `seq` invocation can best be explained by defining them in terms of the `for-loop` as follows. Typically, y is a function of i.

```
seq(y, i = m..n)  ≡  T := NULL; for i from m to n do T := T,y od; T
seq(y, i = x)     ≡  T := NULL; for i in x do T := T,y od; T
```

In either form, the `seq` version is more efficient than the `for-loop` version because the `for-loop` version constructs many intermediate sequences. Specifically, the cost of the `seq` version is linear in the length of the sequence generated but the `for-loop` version is quadratic.

Examples:

```
seq(p[i], i = 1..3);        ⟶     p[1], p[2], p[3]
seq(i^2, i = 2/3..8/3);     ⟶     4/9, 25/9, 64/9
seq(x, i = 1..4);           ⟶     x, x, x, x
x $ 4;                      ⟶     x, x, x, x
seq(i, i = 2..5);           ⟶     2, 3, 4, 5
$ 2..5;                     ⟶     2, 3, 4, 5
L := [seq(i, i = 0..6)];    ⟶     L := [0, 1, 2, 3, 4, 5, 6]
seq(k^2 mod 7, k = L);      ⟶     0, 1, 2, 4
```

As can be noted in the above examples, there is also a *sequence operator* $ for forming expression sequences. In general, it is less efficient than the `seq` function and hence the `seq` function is to be preferred for programming. The general forms of the $ operator are

$$\langle expr \rangle_1 \ \$ \ \langle name \rangle = \langle expr \rangle_2 \ .. \ \langle expr \rangle_3$$
$$\langle expr \rangle_1 \ \$ \ \langle expr \rangle_3$$
$$\$ \ \langle expr \rangle_2 \ .. \ \langle expr \rangle_3 \ .$$

The first form corresponds to the construct

$$\textsf{seq}(\langle expr \rangle_1, \ \langle name \rangle = \langle expr \rangle_2 \ .. \ \langle expr \rangle_3) \ .$$

The second form is a shorthand notation for

$$\langle expr \rangle_1 \ \$ \ i \ = \ 1 \ .. \ \langle expr \rangle_3$$

(assuming that $\langle expr \rangle_1$ does not contain i). In other words, the result of the second form is to produce a sequence of length $\langle expr \rangle_3$ by repetition of $\langle expr \rangle_1$. The third form is a shorthand notation for

$$\texttt{i \$ i} = \langle expr \rangle_2 \; .. \; \langle expr \rangle_3$$

which is a simple mechanism for expanding the range notation into a numeric sequence. In all cases, if either of $\langle expr \rangle_2$ or $\langle expr \rangle_3$ does not evaluate to a numeric value then the result is an unevaluated function invocation of the function named `` `$` `` (but in the usual prettyprinting mode, the output format corresponds to the input format specified above). It should be noted that where a $\langle name \rangle$ is required in the $ construct, it must evaluate to a name; in practice, this may require the use of single quotes to enclose both '$\langle expr \rangle_1$' and '$\langle name \rangle$' to prevent unwanted evaluation (assuming that $\langle name \rangle$ has a previous value). The **seq** function has a more natural semantics since it follows the **for-loop** semantics.

The most common use of the **$** operator (rather than the corresponding **seq** function) is in conjunction with **diff** where the shorthand notation is particularly appealing. For example,

    diff(sin(x*y), x$3, y$2)

is equivalent to `diff(sin(x*y),x,x,x,y,y)` .

3.2.5   Sets and Lists

A *set* is an expression of the form

$$\{ \; \langle expression\ sequence \rangle \; \}$$

and a *list* is an expression of the form

$$[\; \langle expression\ sequence \rangle \;] \; .$$

Note that an $\langle expression\ sequence \rangle$ may be empty so that the empty set is represented by {} and the empty list is represented by []. A set is an *unordered* sequence of expressions and the user should not assume that the expressions will be maintained in any particular order. (A particular invocation of Maple will use a particular ordering based on the internal addresses of the expressions.) Duplicates are removed from the expression sequence. A list is an *ordered* sequence of expressions with the order of the expressions specified by the user. For example, if the user inputs the set {x,y,y} the system might represent this in the form {y,x} whereas if the user inputs the list [x,y,y] then the representation used by the system will be precisely this list.

Some operations which are useful for the manipulation of lists are equality testing via the relational operator '=' (see Section 3.2.8), the *selection* operation (see Section 3.2.10), and the functions **op**, **subsop**, and **map** (see Section 4.2). The particular operation of *list concatentation* is performed by forming a new list, using the **op** function to extract the elements of the given lists.

For example, if L1 := [x,y] and L2 := [y,z,w] then the concatenation of these two lists can be achieved by the operation L := [op(L1), op(L2)] which assigns to L the list [x,y,y,z,w].

There are three *set operators* which are used for set arithmetic:

union, minus, intersect .

As the names imply, these operators perform set union, set difference, and set intersection, respectively. As operators, these three names are reserved words in Maple. It is also possible to use the corresponding back-quoted names:

`union`, `minus`, `intersect`

as function names to achieve the same effect (see Examples below). Some other operations which are useful for the manipulation of sets are equality testing via the relational operator '=' (see Section 3.2.8) and the function **member** (see the Maple Library) which tests for set membership.

Examples:

u := [x,2*x+1,y];	$\longrightarrow$	u := [x, 2*x+1, y]
u[3];	$\longrightarrow$	y
v := [op(2..3, u), exp(x)];	$\longrightarrow$	v := [2*x+1, y, exp(x)]
w := [u[2..3], exp(x)];	$\longrightarrow$	w := [2*x+1, y, exp(x)]
evalb(v=w);	$\longrightarrow$	true
uv := [op(u), op(v)];	$\longrightarrow$	uv := [x, 2*x+1, y, 2*x+1, y, exp(x)]
u1 := [op(u), y^2];	$\longrightarrow$	u1 := [x, 2*x+1, y, y^2]
u2 := subsop(2=z, u1);	$\longrightarrow$	u2 := [x, z, y, y^2]
map(sqrt, u2);	$\longrightarrow$	[x^(1/2), z^(1/2), y^(1/2), y]
s := {x,y} union {y,z};	$\longrightarrow$	s := {x,y,z}
t := `union`({x,y},{y,z});	$\longrightarrow$	t := {x,y,z}
evalb(s=t);	$\longrightarrow$	true
member(y, s);	$\longrightarrow$	true
member(y, {x*y, y*z});	$\longrightarrow$	false
member(x*y, {x*y, y*z});	$\longrightarrow$	true
{3,4} union a union {3,7} union b;	$\longrightarrow$	union(a,b,{3,4,7})
{x,y,z} minus {y,z,w};	$\longrightarrow$	{x}
a minus b;	$\longrightarrow$	a minus b
`minus`(a,a);	$\longrightarrow$	{}
{x,y,z} intersect {y,z,w};	$\longrightarrow$	{y,z}
`intersect`(a,c,b,a);	$\longrightarrow$	intersect(a,b,c)

3.2.6 Algebraic Operators

There are fourteen *algebraic operators*

    ", "", """, !, +, -, *, &*, /, **, ^, @, @@, mod .

The nullary operator " has as its value the latest expression, the nullary operator "" has as its value the penultimate expression, and the nullary operator """ has as its value the expression preceding the penultimate expression. The sequence of expressions defining these three nullary operators is the sequence of expressions generated during a Maple session *at the same level* as the occurrence of the nullary operator, but it specifically excludes the NULL expression sequence and it excludes the values of the index of iteration in a repetition statement. The most common use of these operators is at the top, or interactive, level of a Maple session, in which case the expressions generated by a procedure invocation are "hidden" from these nullary operators because they are at a lower level. These operators may also be used within the body of a procedure, in which case they refer to the sequence of expressions generated during execution of that particular procedure body, excluding expressions generated by any sub-procedures invoked.

The unary operator ! is used as a postfix operator and it denotes the factorial function of its operand. + and – may be used as prefix operators representing unary plus and unary minus. The seven operators

$$+, \quad -, \quad *, \quad \&*, \quad /, \quad **, \quad \hat{ }$$

may all be used as binary operators, representing addition, subtraction, multiplication, noncommutative multiplication, division, exponentiation, and exponentiation, respectively. The two operators ** and ^ are synonymous and may be used interchangeably. The noncommutative multiplication operator &* acts as an inert operator (like the *neutral operators* described in the next section), but the parser understands its binding strength to be equivalent to the binding strength of * and /. The evalm function in the Maple Library interprets &* as the matrix multiplication operator. &* may also be used as a unary operator, and the particular form &*() is interpreted by evalm as the generic identity matrix.

The @ operator is used to represent composition of functional operators. For example,

```
(sin@exp)(x);        ⟶        sin(exp(x))
(sin@arcsin)(x);     ⟶        x
```

The @@ operator is the corresponding "exponentiation" operator representing repeated applications of the composition operator. For example,

```
(ln@@2)(x);              ⟶        ln(ln(x))
((sin@@2)@arcsin)(x);    ⟶        sin(x)
```

For more detailed information on the algebra of functional operators in Maple, see Chapter 8.

The mod operator is used for the operation of evaluating an expression modulo m, for a nonzero integer modulus m. As with the set operators, it is also available as a function with the name `mod`. The representation used for the integers modulo m depends on the specific function assigned to the name `mod`. By default,

```
`mod` := modp
```

which uses the positive representation (thus for an integer n the result of the operation r := n mod m lies in the interval $0 \le r < |m|$). If the user assigns

Content	Det	DistDeg	Divide	Eval
Expand	Factor	Factors	Gcd	Gcdex
Hermite	Interp	Irreduc	Normal	Nullspace
Power	Powmod	Prem	Primitive	Primpart
Quo	Randpoly	Randprime	Rem	Resultant
RootOf	Roots	Smith	Sprem	Sqrfree

TABLE 3.1. Functions known to mod

```
`mod` := mods
```

then the symmetric representation is used (thus for an integer n the result of the operation $r :=$ n mod m lies in the interval $-|\mathtt{m}|/2 < \mathtt{r} \le |\mathtt{m}|/2$). The functions modp and mods may be invoked directly if desired.

The mod operator understands the inert operator &^ for powering. For example,

```
i &^ j mod m
```

will calculate i^j mod m without first calculating i^j in integer arithmetic. It also understands the function name power for this same purpose:

```
power(i,j) mod m  .
```

The first operand of the mod operator is not restricted to be an integer expression but rather it may be a general expression. The expression is evaluated over the coefficient field of integers modulo m. In particular, various functions for polynomial and matrix arithmetic over finite rings and fields are known to the mod operator; these are listed in Table 3.1. For further information on these functions, see the *Maple V Library Reference Manual* or the on-line help pages.

Examples:

```
`mod`;                          ⟶    modp
12 mod 7;                       ⟶    5
7 mod p;                        ⟶    modp(7,p)
`mod` := mods: 12 mod 7;        ⟶    -2
`mod` := modp: -1 mod 13;       ⟶    12
3^(-1) mod 7;                   ⟶    5
17*x*y^11 + 11*x  mod 11;       ⟶    6*x*y^11
2 &^ 8 mod 100;                 ⟶    56
`mod` := mods: 4/3 mod 7;       ⟶    -1
17*x*y^11 + 11*x  mod 11;       ⟶    -5*x*y^11
a := 9*x^3 + 9*x^2 + 6*x + 6:
b := 3*x^2 + 8*x + 9:
g := Gcd(a,b) mod 11;           ⟶    g := x + 5
Divide(a,g,'q') mod 11;         ⟶    true
q;                              ⟶    -2*x^2 - 3*x - 1
```

The order of precedence of all programming-language operators is described in Section 3.2.14 below. However, any expression may be enclosed in parentheses, yielding a new valid expression. This mechanism can be used to force a particular order of evaluation.

3.2.7   Neutral Operators

There is a facility for *user-defined* or *neutral* operators. A neutral operator symbol is formed by the ampersand character '**&**' followed by a sequence of one or more characters. There are two varieties of **&**-names, depending on whether the sequence of characters is alphanumeric or non-alphanumeric.

In the first case, the neutral operator symbol consists of **&** followed by a letter or an underscore, followed by a sequence of zero or more alphanumeric characters (letters, digits, and underscores). In other words, any Maple *string* not requiring back quotes, preceded by the **&** character, forms a valid **&**-name. For this variety of **&**-name, it is necessary to use *white space* to delineate the end of the operator name whenever the succeeding operand is alphanumeric; hence, the non-alphanumeric operator names are generally more convenient to use.

The second variety of **&**-name consists of **&** followed by one or more characters which are non-alphanumeric (thus letters, digits, and underscores are excluded) and which do not appear in the exclusion list below. The exclusion list consists of the ampersand character, all punctuation marks which are not also programming-language operators, the comment character, and the white-space characters. Specifically, the characters excluded from **&**-names (following the **&**) are:

    & | (   )   [   ]   {   }   ;   :   '   `   #

as well as ⟨newline⟩ and ⟨blank⟩. As with strings, the maximum length of an **&**-name is 499 characters.

The particular neutral operator symbol **&*** is singled out as a special token representing the noncommutative multiplication operator. The special property of **&*** is that the parser understands its binding strength to be equivalent to Maple's other multiplication operators. All other neutral operators have binding strength greater than the standard algebraic operators. (See Section 3.2.14 for the order of precedence of all programming-language operators.) The `evalm` function in the Maple Library interprets **&*** as the matrix multiplication operator.

Neutral operators may be used as unary prefix operators, or infix binary operators, or as function calls. They generate function calls, with the name of the function the same as the name of the neutral operator. (In the usual prettyprinting mode, these particular function calls are printed in binary operator format when there are exactly two operands, and in unary operator format when there is exactly one operand, but the internal representation is an unevaluated function.) Maple imposes no semantics on the neutral operators. The user may define manipulations on expressions containing such operators via Maple's interface to user-defined procedures for various standard library functions, including `simplify`, `diff`, `series`, and `evalf`. (See Chapter 8 for a more detailed discussion of operators.)

Examples:

```
a &+ b;            ⟶        a &+ b
&%! a;             ⟶        &%! a
&-(a,b,c,d);       ⟶        &-(a,b,c,d)
a &xx b - &xx(a,b); ⟶       0
a * b &+ c;        ⟶        a*(b &+ c)
d &+ e^2;          ⟶        (d &+ e)^2;
d &* e^2;          ⟶        d &* e^2;
```

3.2.8   Relations and Logical Operators

A new type of expression can be formed from ordinary algebraic expressions by using the *relational operators* <, <=, >, >=, =, <>. The semantics of these operators is dependent on whether they occur in an *algebraic* context or in a *Boolean* context.

In an algebraic context, the relational operators are simply "place holders" for forming equations or inequalities. Addition of equations/inequalities and multiplication of an equation/inequality by an algebraic expression are fully supported in Maple. In the case of adding or subtracting two equations, for example, the addition or subtraction is applied to each side of the equations, thus yielding a new equation. In the case of multiplying an equation by an expression, the multiplication is distributed to each side of the equation. Similar operations may be performed with inequalities.

In a Boolean context, a relation is evaluated to the value **true** or the value **false**. (The names **true** and **false** are defined to be constants in Maple.) In the case of the operators <, <=, >, >= the difference of the operands must evaluate to a numeric constant and this constant is compared with zero. In the case of either of the relations

```
op1 = op2
op1 <> op2
```
the operands may be arbitrary expressions (algebraic or non-algebraic). Note, however, that this equality test for expressions deals only with syntactic equality of the Maple representations of the expressions, which is not the same as mathematical equivalence. For example, the equation

```
x + y  =  y + x
```
will evaluate to **true** in a Boolean context (because the Maple internal representations of these two expressions will be identical). However, the equation

```
x^2 - y^2  =  (x - y) * (x + y)
```
will evaluate to **false** in a Boolean context, even though the expressions are mathematically equivalent. In the latter example, applying the **expand** function results in an equation which will evaluate to **true**.

More generally, an expression can be formed using the *logical operators* **and**, **or**, **not**, where the first two are binary operators and the third is a unary (prefix) operator. An expression containing one or more logical operators is automatically evaluated in a Boolean context. (See Section 3.2.14 regarding the order of evaluation.)

AND	false	true	FAIL
false	false	false	false
true	false	true	FAIL
FAIL	false	FAIL	FAIL

OR	false	true	FAIL
false	false	true	FAIL
true	true	true	true
FAIL	FAIL	true	FAIL

NOT	false	true	FAIL
	true	false	FAIL

TABLE 3.2. Truth Tables

The evaluation of Boolean expressions in Maple uses the following *three-valued logic*. In addition to the special names **true** and **false**, Maple also understands the special name FAIL. The value FAIL is sometimes used as the value returned by a procedure when it is unable to completely solve a problem. In other words, it can be viewed as the value "don't know". In the context of the Boolean clause in an **if**-statement or a **while**-statement, the branching of the program is determined by treating the value FAIL the same as the value **false**. The evaluation of a Boolean expression may yield **true**, **false**, or FAIL according to the following truth tables.

It should be noted that three-valued logic leads to asymmetry in the use of **if**-statements and **while**-statements. For example, the following two statements are not equivalent:

> **if** ⟨condition⟩ **then** ⟨statseq1⟩ **else** ⟨statseq2⟩ **fi**
> **if not** ⟨condition⟩ **then** ⟨statseq2⟩ **else** ⟨statseq1⟩ **fi** .

Depending on the desired action in the case where ⟨condition⟩ has the value FAIL, either the first or the second of these two **if**-statements may be correct for a particular context.

3.2.9 Ranges

Yet another type of expression is a *range* which is formed using the ellipsis operator:

> ⟨expression⟩ .. ⟨expression⟩

where the operator here can be specified as two *or more* consecutive periods. Any valid Maple ⟨expression⟩s may appear as the operands. The ellipsis operator simply acts as a "place holder" in the same manner as when the relational operators are used in an algebraic context. To convert a range into an expression sequence, the $ operator may be applied (see Section 3.2.4 above). For example:

```
1..5;           ⟶        1 .. 5
$1..5;          ⟶        1, 2, 3, 4, 5
```

Two common uses of ranges are for Maple's library functions **sum** and **int**. For example, in the function call

```
sum( i^2, i = 1 .. n )
```

the sum function interprets this to mean that the lower and upper limits of summation are 1 and n, respectively. Similarly, in the function call

```
int( exp(2*x), x = 3/4 .. 7/8 )
```

the integration function interprets this as a definite integration with lower and upper limits of integration 3/4 and 7/8, respectively. The range construct is also used by Maple's built-in function **op**, which extracts operands from an expression. For example,

```
a := [ x, y, z, w ]; ⟶      a := [ x, y, z, w ]
op(2, a);            ⟶      y
op(3, a);            ⟶      z
op(2..4, a);         ⟶      y, z, w .
```

The ⟨range⟩ construct can also be used in combination with the concatenation operator to form an ⟨expression sequence⟩ as follows. The construct

$$\langle name \rangle \ . \ (\ \langle range \rangle \ )$$

is a generalization of name formation using the concatenation operator. Here, the left and right operands in ⟨range⟩ must be integers, and the result is an ⟨expression sequence⟩ consisting of the names formed by concatenating ⟨name⟩ to each integer in the specified range. If the right operand in ⟨range⟩ is less than the left operand then the result is the null expression sequence. For example,

```
p . (1 .. 5)        ⟶        p1, p2, p3, p4, p5
```

3.2.10 Selection Operation

The selection operation '[ ]' can be used to select components from an aggregate object. The syntax for the selection operation is

$$\langle name \rangle \ [\ \langle expression \ sequence \rangle \ ] \ .$$

The ⟨name⟩ must evaluate to one of the following types:

- **name**

- **table**

- `array`

- `list`

- `set`

- `expression sequence` .

If ⟨name⟩ evaluates to an unassigned **name** then

$$⟨name⟩ \ [\ ⟨expression \ sequence⟩ \ ]$$

is an indexed name as described earlier (see Section 3.2.2). If ⟨name⟩ evaluates to a **table** or an **array** then the selection operation is an indexing operation. The use of arrays, tables, and indexing operations is discussed in Chapter 6. If ⟨name⟩ evaluates to a **list**, **set**, or **expression sequence** then ⟨expression sequence⟩ must evaluate to an integer, a range, or NULL. If ⟨expression sequence⟩ is an integer i then the ith operand of the aggregate object is returned. If ⟨expression sequence⟩ is a range then an expression sequence is returned containing the operands of the aggregate object as specified by the range. If ⟨expression sequence⟩ is empty then an expression sequence containing all of the operands of the aggregate object is returned. Thus there is the following explicit relationship between the selection operation and the **op** function (see the Maple Library) in the case where ⟨name⟩ evaluates to a list or a set:

$$⟨name⟩ \ [\ ⟨expression \ sequence⟩ \ ]$$

is equivalent to

$$op(\ ⟨expression \ sequence⟩ \ , \ ⟨name⟩ \ ).$$

For example,

```
a := [w,x,y,z];      ⟶      a := [w,x,y,z]
a[3];                ⟶      y
a[1..3];             ⟶      w, x, y
a[ ];                ⟶      w, x, y, z
```

3.2.11 Unevaluated Expressions

An expression enclosed in a pair of single quotes is called an *unevaluated expression*. For example, the statements

```
a := 1;  x := a + b;
```

cause the value 1+b to be assigned to the name x, while the statements

```
a := 1;  x := 'a + b';
```

cause the value **a+b** to be assigned to the name **x**. The latter effect can also be achieved (if **b** has no value) by the statements

```
a := 1;   x := 'a' + b;
```

The effect of evaluating a quoted expression is to strip off (one level of) quotes, so in some cases it is useful to use nested levels of quotes. Note that there is a distinction between *evaluation* and *simplification* so that the statement

```
x := '2 + 3';
```

will cause the value 5 to be assigned to the name **x** even though the expression appearing here is quoted. The evaluator simply strips off the quotes, but it is the *simplifier* which transforms the expression 2 + 3 into the constant 5.

A special case of "unevaluation" arises when a name which may have been assigned a value needs to be unassigned, so that in the future the name simply stands for itself. This is accomplished by assigning the quoted name to itself. For example, if the statement

```
x := 'x';
```

is executed, then even if **x** had previously been assigned a value it will now stand for itself in the same manner as if it had never been assigned a value. (See also the function **evaln** in the Maple Library.)

3.2.12 Procedures and Functions

Another valid expression in Maple is a *procedure definition* which takes the form

> **proc** (⟨nameseq⟩) **local** ⟨nameseq⟩; **options** ⟨nameseq⟩; ⟨statseq⟩ **end**

where the '**local** ⟨nameseq⟩;' part and the '**options** ⟨nameseq⟩;' part may be omitted. Here, ⟨nameseq⟩ stands for a sequence of ⟨name⟩s (specifically, they must be sequences of ⟨string⟩s) and in the first instance above, the ⟨nameseq⟩ may be empty. This construct has some similarities with the concept of unevaluated expressions, but in this case it is more generally a ⟨statseq⟩ (a sequence of statements) which is unevaluated. Note that the keywords **proc** and **end** serve a purpose similar to the single quotes in unevaluated expressions, although evaluation of this expression does not cause these keywords to be stripped off. An example of a procedure definition is

```
max := proc ( a, b ) if a>b then a else b fi end
```

which is syntactically an assignment statement where the ⟨expression⟩ on the right-hand side is a procedure definition.

A procedure is invoked by a *procedure invocation* which normally has the syntax

> ⟨name⟩ (⟨expression sequence⟩) .

For example, if **max** is defined as above then the expression **max(1,2)** is a procedure invocation in which the *actual parameters* 1 and 2 are substituted for the *formal parameters* **a** and **b**, respectively, and then the procedure body is executed, yielding the value 2 in this case. For a more general discussion of procedures, see Chapter 7.

The above syntax for a procedure invocation may also be used in cases where the ⟨name⟩ has not been assigned, in which case the result is an *unevaluated function*, which is another instance of a Maple expression. Examples of unevaluated functions are: `sin(x)`, `exp(x^2)`, and `f(x,y)`.

More generally, a procedure invocation may take the form:

⟨expr⟩ (⟨expression sequence⟩)

where ⟨expr⟩ may be any of the following types of expressions:

- name

- procedure invocation

- procedure definition

- functional operator

- integer

- float

- parenthesized algebraic expression .

For example, the library function `psqrt` is not *readlib-defined* (see Chapter 12) so it must be explicitly loaded using `readlib(psqrt)`. Rather than the pair of statements

```
readlib(psqrt):
p := psqrt(x^2+2*x+1);
```
it is possible to use the single statement
```
p := readlib(psqrt)(x^2+2*x+1);
```
to achieve the same action. An example of using a procedure definition directly in a procedure invocation is
```
 := proc(x) x^2 end (5);a
```
which assigns the value 25 to **a**. The other cases listed above are discussed in the context of Maple's algebra of functional operators (see Section 3.2.13 below and Chapter 8).

3.2.13   Functional Operators

A functional operator in Maple is a special form of a procedure. There are two equivalent notations for functional operators: the *arrow* notation

(⟨vars⟩) -> ⟨result⟩

and the *angle-bracket* notation

$$< \langle \text{result} \rangle \ | \ \langle \text{vars} \rangle > \ .$$

Here, $\langle \text{vars} \rangle$ is a sequence of variable names (or a single variable) and $\langle \text{result} \rangle$ is the result of the procedure acting on $\langle \text{vars} \rangle$. For example, in arrow notation,

```
x -> x^2
```

represents the function that squares its argument. In angle-bracket notation, this same function would be written

```
< x^2 | x > .
```

Multivariate and vector functions are also allowed. When using arrow notation, it is necessary to put parentheses around $\langle \text{vars} \rangle$ or $\langle \text{result} \rangle$ whenever they are expression sequences. When using angle-bracket notation, $\langle \text{result} \rangle$ must be parenthesized if it is an expression sequence but $\langle \text{vars} \rangle$ must not be parenthesized. For example, the following functions have the correct syntax:

```
(x,y) -> x^2 + y^2      or equivalently   < x^2 + y^2 | x,y >
x -> (2*x, 3*x^4)       or equivalently   < (2*x, 3*x^4) | x >
(x,y,z) -> (x*y, y*z)   or equivalently   < (x*y, y*z) | x,y,z > .
```

Another way to create a functional operator is by using the **unapply** function. (See **unapply** in the Maple Library.)

Any of the following four forms of angle-bracket notation may be used to form a functional operator from an expression $\langle \text{result} \rangle$:

$$< \langle \text{result} \rangle >$$
$$< \langle \text{result} \rangle \ | \ \langle \text{vars} \rangle >$$
$$< \langle \text{result} \rangle \ | \ \langle \text{vars} \rangle \ | \ \langle \text{localvars} \rangle >$$
$$< \langle \text{result} \rangle \ || \ \langle \text{localvars} \rangle >.$$

If the first alternative is used then $\langle \text{vars} \rangle$ is automatically determined to be the sequence of all names appearing in $\langle \text{result} \rangle$ which are not system constants. Note that this shorthand notation is mainly useful in the case of a single argument, because for multiple arguments the user will have no control over the order of the $\langle \text{vars} \rangle$. This angle-bracket notation is then equivalent to

$$(\ \langle \text{vars} \rangle \ ) \ -> \langle \text{result} \rangle \ .$$

The second alternative is equivalent to

$$(\ \langle \text{vars} \rangle \ ) \ -> \langle \text{result} \rangle \ ,$$

that is, a function which takes $\langle \text{vars} \rangle$ as its arguments and returns $\langle \text{result} \rangle$. The third alternative is to be used when local variables are needed inside the procedure. It is equivalent to the arrow notation

$$(\langle \text{vars} \rangle) \to \textbf{local} \langle \text{localvars} \rangle ; \langle \text{result} \rangle \, .$$

For example, the function

```
< int(sin(x), x=a..b) | a,b | x >
```

takes two arguments, a and b, and returns the definite integral of the sin function from a to b. The variable x is just a dummy variable and hence we do not want the function definition to be affected by any global value x that might be assigned. The equivalent function in arrow notation is

```
(a,b) -> local x; int(sin(x), x=a..b) .
```

The final alternative is used to create a function with no arguments but with local variables. The equivalent arrow notation is

$$(\,) \to \textbf{local} \langle \text{localvars} \rangle ; \langle \text{result} \rangle \, .$$

The semantics of functional operators can best be thought of as a process of *procedurizing* the given expression **result**, since the internal data structure representation is a procedure. Referring to the four cases discussed above in angle-bracket notation, in the first two cases the procedure definition is

$$\textbf{proc} \, (\, \langle \text{vars} \rangle \,) \, \textbf{options} \, \texttt{operator, angle;} \, \langle \text{result} \rangle \, \textbf{end}$$

where in the first case, $\langle \text{vars} \rangle$ has been deduced from the names appearing in **result**. In the third case, the representation is

$$\textbf{proc} \, (\, \langle \text{vars} \rangle \,) \, \textbf{local} \langle \text{localvars} \rangle ; \, \textbf{options} \, \texttt{operator, angle;} \, \langle \text{result} \rangle \, \textbf{end}$$

and in the fourth case

$$\textbf{proc} \, (\,) \, \textbf{local} \langle \text{localvars} \rangle ; \, \textbf{options} \, \texttt{operator, angle;} \, \langle \text{result} \rangle \, \textbf{end} \, .$$

When the functional operator is expressed using arrow notation, the procedure definition is as above except that the options field becomes `options operator, arrow` . It follows that one can switch between angle-bracket and arrow notations via the command `subs('angle' = 'arrow', eval(f))` switching f from angle to arrow, or `subs('arrow' = 'angle', eval(f))` switching f from arrow to angle).

Maple performs some automatic simplifications on functional operators (more precisely, on procedures with option `operator`). For example,

```
F := x -> sin(x);        ⟶        F := sin
```

where the functional operator has been simplified to the function name sin. The reason is that the action of applying the operator F to an argument will be precisely the action of applying the function name sin to the argument. Further examples are:

```
A := <1>;                ⟶        A := 1
B := <3.14>;             ⟶        B := 3.14
```

which are examples of *constant operators*. The result of applying a constant to some arguments in a functional notation is the constant itself:

`<1>(x);`	$\longrightarrow$	`1`
`1(x);`	$\longrightarrow$	`1`
`<3.14>(x,y);`	$\longrightarrow$	`3.14`
`3.14(x,y);`	$\longrightarrow$	`3.14`

Note that the *identity operator* is expressed as `x -> x` or in angle-bracket notation, `<x>` for any choice of parameter name `x`. Note also that parenthesized algebraic expressions, when applied to some arguments in a functional notation, are interpreted in an "operator algebra".

Examples:

`<x>(t);`	$\longrightarrow$	`t`
`(a+b)(t);`	$\longrightarrow$	`a(t) + b(t)`
`(a+1)(t);`	$\longrightarrow$	`a(t) + 1`
`( <ln(x)+1>@@2 )(t);`	$\longrightarrow$	`ln(ln(t) + 1) + 1`
`( <sin(x)>@<arcsin(x)> )(t);`	$\longrightarrow$	`t`

For a more detailed discussion of functional operators, see Chapter 8.

3.2.14 Precedence of Programming-Language Operators

The order of precedence of all unary and binary operators is listed in the following table, from highest to lowest binding strengths. In parentheses it is stated whether the operators are left associative, right associative, or non-associative. However, note that any expression may be enclosed in parentheses, yielding a new valid expression, and this mechanism can be used to force a particular order of evaluation. Thus the concatenation or decimal point operator '.' has the highest binding strength and the assignment operator ':=' has the lowest binding strength. Note that the exponentiation operators `^`, `**`, and `@@` are defined to be non-associative and therefore `a^b^c` is syntactically invalid in Maple. (The user must use parentheses to state the desired interpretation explicitly.)

The evaluation of expressions involving the logical operators proceeds in an intelligent manner which exploits more than the simple associativity and precedence of these operators. Namely, the left operand of the operators **and** and **or** is always evaluated first and the evaluation of the right operand is avoided if the truth value of the expression can be deduced from the value of the left operand alone. For example, the construct

```
if  d <> 0  and  f(d)/d > 1  then . . . fi
```

will not cause a division by zero, because if `d=0` then the left operand of **and** becomes **false** and the right operand of **and** will not be evaluated.

.	(left associative)
%	(non-associative)
&-operators	(left associative)
!	(left associative)
^, **, @@	(non-associative)
, &, /, @, intersect	(left associative)
+, -, union, minus	(left associative)
mod	(non-associative)
..	(non-associative)
<, <=, >, >=, =, <>	(non-associative)
$	(non-associative)
not	(right associative)
and	(left associative)
or	(left associative)
->	(right associative)
,	(left associative)
:=	(non-associative)

TABLE 3.3. Order of precedence of operators

3.3   Formal Syntax

This section presents the BNF grammar which describes the syntax accepted by Maple. In the following grammar, where a sequence of symbols is enclosed in a pair of "‖" symbols it indicates that this portion of the statement is optional. Where *empty* occurs in the grammar, no symbol is required. A Maple *session* consists of a ⟨statseq⟩, which is a sequence of statements separated by semicolons or colons. At the top (interactive) level of the Maple system, the colon serves to suppress the display of the result of the preceding statement. Within Maple procedures, the colon and the semicolon are interchangeable.

⟨statseq⟩ ::=   ⟨statseq⟩ ; ⟨stat⟩ | ⟨statseq⟩ : ⟨stat⟩
               | ⟨stat⟩

⟨stat⟩ ::=   ⟨name⟩ := ⟨expr⟩
              | ⟨expr⟩
              | **read** ⟨expr⟩
              | **save** ⟨expr⟩
              | ⟨ifpart⟩ **fi**
              | ⟨ifpart⟩ **else** ⟨statseq⟩ **fi**
              | ‖ **for** ⟨name⟩ ‖ ‖ **from** ⟨expr⟩ ‖ ‖ **by** ⟨expr⟩ ‖
                ‖ **to** ⟨expr⟩ ‖ ‖ **while** ⟨expr⟩ ‖
                  do ⟨statseq⟩ **od**
              | ‖ **for** ⟨name⟩ ‖ ‖ **in** ⟨expr⟩ ‖ ‖ **while** ⟨expr⟩ ‖
                  do ⟨statseq⟩ **od**
              | **quit**[2]
              | *empty*

⟨ifpart⟩ ::=   **if** ⟨expr⟩ **then** ⟨statseq⟩
              | ⟨ifpart⟩ **elif** ⟨expr⟩ **then** ⟨statseq⟩

⟨expr⟩ ::=   ⟨expr⟩ , ⟨expr⟩   *expression sequence*

              | ⟨name⟩ -> ‖ **local** ⟨nameseq⟩ ; ‖ ⟨expr⟩   *arrow operators*
              | (⟨exprseq⟩) -> ‖ **local** ⟨nameseq⟩ ; ‖ ⟨expr⟩

              | ⟨expr⟩ **or** ⟨expr⟩   *boolean expressions*
              | ⟨expr⟩ **and** ⟨expr⟩ | **not** ⟨expr⟩

              | ⟨expr⟩ $ ⟨expr⟩ | $ ⟨expr⟩   *sequence generators*

              | ⟨expr⟩ < ⟨expr⟩ | ⟨expr⟩ <= ⟨expr⟩   *relations*
              | ⟨expr⟩ > ⟨expr⟩ | ⟨expr⟩ >= ⟨expr⟩
              | ⟨expr⟩ = ⟨expr⟩ | ⟨expr⟩ <> ⟨expr⟩

              | ⟨expr⟩ ..[3] ⟨expr⟩   *range*

| ⟨expr⟩ mod ⟨expr⟩   *modulo operation*

| + ⟨expr⟩   *algebraic and set operations*
| − ⟨expr⟩
| ⟨expr⟩ + ⟨expr⟩ | ⟨expr⟩ − ⟨expr⟩
| ⟨expr⟩ union ⟨expr⟩ | ⟨expr⟩ minus ⟨expr⟩
| ⟨expr⟩ * ⟨expr⟩ | ⟨expr⟩ / ⟨expr⟩
| ⟨expr⟩ &* ⟨expr⟩ | ⟨expr⟩ intersect ⟨expr⟩
| ⟨expr⟩ @ ⟨expr⟩ | ⟨expr⟩ @@ ⟨expr⟩
| ⟨expr⟩ ^ ⟨expr⟩ | ⟨expr⟩ ** ⟨expr⟩

| ⟨expr⟩ !   *factorial*

| ⟨expr⟩ ⟨neutralop⟩ ⟨expr⟩   *neutral operators*
| ⟨neutralop⟩ ⟨expr⟩
| &* ⟨expr⟩

| **proc** (‖ ⟨nameseq⟩ ‖)   *procedure definition*
    ‖ **local** ⟨nameseq⟩ ; ‖
    ‖ **options** ⟨nameseq⟩ ; ‖ [4]
    ⟨statseq⟩
  end

| ⟨name⟩   *variable name*
| (⟨expr⟩)   *parenthesized expression*
| ' ⟨expr⟩ '   *unevaluated expression*

| < ⟨expr⟩ >   *functional operators*
| < ⟨expr⟩ | ⟨nameseq⟩ >
| < ⟨expr⟩ | ‖ ⟨nameseq⟩ ‖ | ⟨nameseq⟩ >

| ⟨expr⟩ (⟨exprseq⟩)   *function call*
| ⟨neutralop⟩ (⟨exprseq⟩)
| &* (⟨exprseq⟩)

| ⟨natural⟩ . ⟨natural⟩   *floating-point numbers*

| ⟨natural⟩ .
| . ⟨natural⟩

| ⟨natural⟩   *unsigned integer*
| { ⟨exprseq⟩ }   *set*
| [⟨exprseq⟩]   *list*

| % ⟨natural⟩   *label*

| "   *previous expressions*
| " " | " " "

⟨exprseq⟩ ::=   ⟨expr⟩ | *empty*

⟨nameseq⟩ ::=   ⟨nameseq⟩ , ⟨string⟩ | ⟨string⟩

⟨name⟩ ::=   ⟨string⟩ | ⟨name⟩ . ⟨natural⟩
            | ⟨name⟩ . ⟨string⟩ | ⟨name⟩ . (⟨expr⟩)
            | ⟨name⟩ [⟨exprseq⟩]
            | ⟨name⟩ (⟨exprseq⟩)

⟨string⟩ ::=   ⟨simple string⟩ | ⟨quoted string⟩   *lexical tokens*

⟨simple string⟩ ::=   *A sequence of one or more letters,*
                       digits, and underscores beginning with
                       either a letter or underscore

⟨quoted string⟩ ::=   *Any sequence of characters in the*
                       supported character set enclosed within
                       a pair of back quotes ``

⟨neutralop⟩ ::=   ⟨neutralop1⟩ | ⟨neutralop2⟩

⟨neutralop1⟩ ::=   *The ampersand character* & *followed by a*
                       ⟨simple string⟩

⟨neutralop2⟩ ::=   *The ampersand character & followed by*
one or more characters excluding: letters,
digits, underscore, & | ( ) [ ] { }
; : ' ` # newline blank

⟨natural⟩ ::=   *A sequence of one or more digits*

---

[2]**done** or **stop** may be used as synonyms for **quit** . In many implementations, pressing the break/interrupt key (or some similar system-dependent sequence) twice in rapid succession will also cause Maple to exit.

[3]Actually, two *or more* consecutive periods are permitted.

[4]**option** may be used as a synonym for **options**.

4
Data Types

4.1 Basic Data Types

Every expression in Maple is represented internally by an expression tree where each node is a particular data type. While some data types are strictly internal use, most of the data types corresponding to expressions are accessible to the user and can be tested for via the **type** function. The user can examine the components of such a data type by using the **op** function, and the corresponding **nops** function can be used to query the *number of operands* of an expression. In this section we discuss the data types which are accessible to the user. For a more detailed description of the internal data types and their representation, see Chapter 9.

The syntax and functionality of the **type**, **op**, and **nops** functions are described in the *Maple V Library Reference Manual* and in on-line help pages. Since the following descriptions make frequent references to the **op** function, the following details should be noted. There are three forms used:

    op(i, expr);   op(i..j, expr);   op(expr);

where i and j must be nonnegative integers. The first form is used to extract the ith operand from **expr**. Typically i is greater than or equal to one, but for some data types a $zero^{th}$ operand is defined as noted in the following subsections. The second form is used to extract the ith through j^{th} operands from **expr**, yielding an expression sequence containing j-i+1 subexpressions, and the result is the null expression sequence if j < i. The third form is equivalent to **op(1..n, expr)** where n is the number of operands appearing in **expr** (ignoring the $zero^{th}$ operand if it is defined).

4.1.1 Integer

An expression is of type **integer** if it is an (optionally signed) sequence of one or more digits of arbitrary length. The length limit is system-dependent but generally much larger than users will encounter — typically greater than 500,000 decimal digits. The **op** function considers this data type to have only one operand, so if **n** is an integer then the value of **op(n)**, and also the value of **op(1, n)**, is the integer n.

4.1.2   Fraction

An expression of type `fraction` is represented by a pair of integers (numerator and denominator) with all common factors removed and with a positive denominator. Like integers, fractions are of arbitrary length. The `op` function considers this data type to have two operands, where the first operand is the numerator and the second operand is the denominator.

The `type` function also understands the composite type `rational`, which is defined to be either an `integer` or a `fraction`.

4.1.3   Floating-Point Number

A floating-point number (called type `float`) is represented externally as a sequence of digits with a decimal point, possibly multiplied by a power of ten. For example,

```
1.5, 15000., .15, .1500000000*10^15    .
```
Floating-point numbers are represented internally by a pair of integers, the mantissa and the exponent, which represent the number

$$\text{mantissa} \times 10^{\text{exponent}}.$$

The exponent in this representation is restricted to be a *word-size* integer on the host computer. (This size limit is system-dependent but typically is nine or ten digits in length.) Thus the `op` function considers this data type to have two operands. For example, `op(150.1)` yields the expression sequence `1501, -1`.

For arithmetic operations and invocations of standard functions, if one of the operands is a floating-point number then floating-point evaluation will take place automatically. More generally, the `evalf` function may be used to force evaluation to a floating-point number. The number of digits carried in the `mantissa` when evaluating floating-point numbers is determined by the value of the global name `Digits` which has `10` as its initial value.

Alternatively, the format `Float(mantissa, exponent)` can be used; it corresponds directly to Maple's internal data structure representation of floating-point numbers, where both arguments to the `Float` function must be integers. When Maple's line-printing mode is used, via the `lprint` function or in normal display when pretty-printing is turned off, the `Float(mantissa, exponent)` notation is used for floating-point numbers with very small or very large magnitudes. For example,

```
interface(prettyprint = false);  evalf(exp(-10));  evalf(exp(-20));
```
yields

```
Float(4539992976,-14)
Float(2061153622,-18)
```
in contrast with the normal display of these numbers which would be seen in pretty-printed format as

```
.00004539992976
```

$$.2061153622*10^{-8}.$$

The **type** function also understands the composite type **numeric**, which is defined to be an **integer**, a **fraction**, or a **float**.

4.1.4 String

An expression is of type **string** if it is either a *simple string* or a *quoted string*. Specifically, a *simple string* is a sequence of one or more letters, digits, and underscores beginning with either a letter or an underscore; a *quoted string* is any sequence of characters in the supported character set formed by enclosing the characters within a pair of back quotes. The **type** function understands the composite type **name**, which is defined to be a **string** or an **indexed** name.

The **string** data type has only one operand, which is the value assigned to it in its role as a name, or else its own name if no other value has been assigned. For example, if x has not been assigned a value then x is of type **string** and **op(x)** has the value x. Another example of a Maple string is `This is a string`. The maximum length of a string is 499 characters.

Note that the construct **op('x')** serves as a *one-level evaluation* operation in Maple, for any string x. This is because the argument will evaluate to the string x, rather than its value, and since the *operand* of a string is defined to be the value assigned to it. In other words, this construct yields the value "pointed at" by the string (name) x without further evaluation of that value. For example,

x := y:	⟶	x := y
y := 2/3:	⟶	y := 2/3
x;	⟶	2/3
op('x');	⟶	y
op(x);	⟶	2, 3

Normal evaluation of x (meaning full recursive evaluation) yields the value 2/3, but the **op** function with argument **'x'** (unevaluated x) yields the value immediately assigned to x. In the last statement above, the argument to the **op** function is the value of x, namely 2/3, which has two operands. Note that another mechanism to achieve *one-level evaluation* is the construct **eval(x, 1)**. (See the **eval** function in the Maple Library.)

4.1.5 Indexed Name

An indexed name, called type **indexed**, is of the form

⟨name⟩ [⟨expression sequence⟩]

where ⟨name⟩ is an unassigned name. The ⟨name⟩ could be a string, or it could be an indexed name itself so there can be a succession of subscripts, as in

```
A[1,2,3][x,y][2,1]    .
```

The op function applied to this data type will yield the expression sequence of indices appearing within the rightmost pair of brackets, and the zeroth operand is the corresponding ⟨name⟩. For example, if expr evaluates to the above indexed name then

```
op( 0, expr );          ⟶          A[1,2,3][x,y]
op( 1, expr );          ⟶          2
op( 2, expr );          ⟶          1
```

As mentioned in the preceding subsection, the type function understands the composite type name which is defined to be a string or an indexed name.

4.1.6 Addition, Multiplication, and Exponentiation

An expression can be composed using the algebraic operators +, -, *, /, ^ (or **). Such an expression is of type `+`, `*`, or `^` (equivalently `**`). Thus the expression a-b is of type `+` and op(a-b) yields the expression sequence a, -b. Similarly the expression a/b is of type `*` and op(a/b) yields the expression sequence a, 1/b. The expression 1/b is an example of an expression of type '^' since it is represented internally as b^(-1). The representation used for these algebraic expressions is often referred to as sum-of-products form.

An expression of type '^' always has exactly two operands, the base expression and the exponent expression. An expression of either type `+` or type `*` can have two or more operands. For example,

```
op( x+y+z+w );        ⟶          x, y, z, w
op( 2*x^2*y );        ⟶          2, x^2, y
op( (x+y)^z );        ⟶          x + y, z .
```

4.1.7 Series

The series data type in Maple is a special data type which represents an expression as a (truncated) power series with respect to a specified indeterminate, expanded about a particular point. This data type is created by a call to the series function. For this data type, the zeroth operand is x - a where x denotes the specified indeterminate and a denotes the particular point of expansion. The first, third, . . . operands are the coefficients (general expressions) and the second, fourth, . . . operands are the corresponding integer exponents. The exponents are restricted to *word-size* integers on the host computer, with a typical length limit of nine or ten digits, ordered from least to greatest. Usually, the final pair of operands in this data type are the special *order* symbol O(1) and the integer n which indicates the order of truncation. (Note: The print routine displays the final pair of operands using the notation $O(x^n)$ rather than more directly as $O(1)x^n$, where x represents the zeroth operand.) However, if the series is known to be exact then there will be no order term in the series. An example of this occurs when the series function is applied to a polynomial whose degree is less than the truncation degree for the series. A very special case is the *zero series*, which is immediately simplified to the integer zero in Maple.

The power series represented by this data structure are generalized power series, which include Laurent series with finite principal parts. More generally, the series coefficients are allowed to depend on **x** under the following restriction. Formally, if the coefficients c_i are dependent on **x** then for any $\epsilon > 0$

$$k_1(\mathbf{x} - a)^\epsilon < |c_i| < k_2(\mathbf{x} - a)^{-\epsilon}, \quad \forall\, \mathbf{x}, 0 < \mathbf{x} - a < r$$

for some positive constants k_1, k_2, and r. In other words, the coefficients may depend on **x** but their growth must be less than polynomial in **x**. `O(1)` represents such a coefficient, rather than an arbitrary constant. An example of a non-standard generalized power series is

$$x^x = 1 + \ln(x)x + \frac{\ln(x)^2 x^2}{2} + \frac{\ln(x)^3 x^3}{6} + \frac{\ln(x)^4 x^4}{24} + O(x^5)\,.$$

The **type** function also understands the composite type **algebraic**, which corresponds to all data types discussed in Sections 4.1.1 through 4.1.7, plus type **function** (Section 4.1.15), type `` `.` `` (Section 4.1.16) and type **uneval** (Section 4.1.17).

4.1.8 Relation

An expression of type **equation** (also called type `` `=` ``) has two operands, the left-hand expression and the right-hand expression. The functions **lhs** and **rhs** can be used as alternative to **op**. For example,

```
e := a = b^3+1;       ⟶        e := a = b^3 + 1
nops(e);              ⟶        2
op(1,e);              ⟶        a
op(1,e);              ⟶        a
lhs(e);               ⟶        a
rhs(e);               ⟶        b^3 + 1
```

An equation is represented externally using the binary operator =.

There are three internal data types for inequalities, corresponding to the operators <>, < and <=. Inequalities involving the operators > and >= are converted to the latter two cases for purposes of representation. Correspondingly, only three names are known to the **type** function for inequalities: `` `<>` ``, `` `<` ``, `` `<=` `` . Like an equation an inequality has two operands, the left-hand-side expression and the right-hand-side expression.

The **type** function also understands the composite type **relation**, which is defined to be one of: `` `=` ``, `` `<>` ``, `` `<` ``, `` `<=` `` .

4.1.9 Boolean Expression

The simplest cases of Boolean expressions are the names `true`, `false`, and `FAIL`. These names are defined to be constants in Maple.

Equations and inequalities (formed using the relational operators =, <>, <, <=, >, >=) are also treated as Boolean expressions if they appear in a *Boolean context*. More complicated boolean expressions can be built out of these simple expressions with the logical operators **and**, **or**, and **not**. The built-in function `evalb` can be called with a Boolean expression as an argument in order to cause the expression to be evaluated as a Boolean. For example, the equation `a=b` is an algebraic equation if it appears alone but `evalb(a=b)` will evaluate this equation as a Boolean. However, an equation or inequality will be recognized as being in a Boolean context if it appears in the `while`-part of a repetition statement, in the `if`-part of a selection statement, or if it is an operand in an expression formed using one or more of the logical operators **and**, **or**, or **not**.

The evaluation of Boolean expressions in Maple uses *three-valued logic* (see Section 3.2.8). In addition to the special names `true` and `false`, Maple also understands the special name `FAIL` which is sometimes used as the value returned from a procedure when a computation is not completely successful.

In addition to the type names for relations, the following type names are known to the **type** function: `` `and` ``, `` `or` ``, `` `not` ``. Each of the first two types of expressions has exactly two operands which can be selected by the **op** function, and the third type has exactly one operand. For example:

```
op( a and b and c );      ⟶    a and b, c
op( a or b and c );       ⟶    a, b and c
op( not(a and b) );       ⟶    a and b .
```

The **type** function also understands the composite type **boolean**, which is defined to be one of the names `true`, `false`, or `FAIL`, or one of the types: `` `and` ``, `` `or` ``, `` `not` ``, `` `=` ``, `` `<>` ``, `` `<` ``, `` `<=` ``.

4.1.10 Range

An expression of type **range** (also called type `` `..` ``) has two operands, the left-hand expression and the right-hand expression. The functions `lhs` and `rhs` can be used as alternative to **op**. For example,

```
r := -2..2;        ⟶    r := -2 .. 2
nops(r);           ⟶    2
op(1,r);           ⟶    -2
op(2,r);           ⟶    2
lhs(r);            ⟶    -2
rhs(r);            ⟶    2
```

A range is represented externally using the binary operator `..` which simply acts as a placeholder in the same manner as the relational operators. For example, the range `1..3` is not equivalent to the expression sequence 1, 2, 3. In certain contexts it is understood to represent this sequence, such as in the construct p.(1..3) or in the construct sum(i, i = 1..3). However, if the actual expression sequence is desired the sequence operator $ (see Section 3.2.4) can be used in the form $1..3 .

4.1.11   Expression Sequence

There is a data type, called type exprseq, for an ⟨expression sequence⟩, which is a sequence of ⟨expression⟩s separated by commas. The operands appearing in an expression sequence cannot be directly extracted via the op function due to the nature of an expression sequence. For example,

```
s := x, y, z;
op(2, s);
```
yields
```
Error, wrong number (or type) of parameters in function op;
```
This is because the arguments to the op function are specified by the expression sequence 2, x, y, z which is too many arguments for the op function. If it is desired to extract the second operand from the expression sequence s, this can be accomplished via op if the expression sequence is placed into a list:
```
op(2, [s])            ⟶        y .
```
Alternatively, the *selection operation* discussed in Section 3.2.10 may be used without forming a list:
```
s[2];                 ⟶        y .
```
When the op function is used to extract parts of an expression, the result is often an expression sequence. For example,
```
a := [x,y,z,w];  op(a);
```
yields the expression sequence x, y, z, w. An important special case of an expression sequence is the null expression sequence and there is a global name in Maple, NULL, whose value is the null expression sequence. The value of the global name NULL is equivalent to the value of the operation op([ ]).

4.1.12   Set and List

Two more data types are the set and the list. Each of these types consists of a sequence of expressions and if expr is an object of either of these two types then op(expr) yields the expression sequence. The external representation of a set uses braces {, }to surround the expression sequence, while the external representation of a list uses brackets [,]to surround the expression sequence. The empty set is represented by { } and the empty list by [ ].

4.1.13   Table and Array

The `table` data type in Maple is created either explicitly via the `table` function call or implicitly by assignment to an indexed name. For example, the statement

```
b := table( [(1)=x] );
```

creates a table object with one component and assigns the table object to the name `b`, while the following statement has precisely the same effect if `b` has no previous value:

```
b[1] := x;
```

The `array` data type in Maple is a specialization of the `table` data type. Specifically, a Maple array is a table with specified dimensions with each dimension an integer range. An array is created via the `array` function call. An array object consists of three parts: an indexing function, an index bound, and a collection of components. Each of these three parts can be selected by the `op` function: the first operand is the indexing function, the second operand is the index bound, and the third operand is the collection of components specified as a list of equations. Either of the first two operands may be `NULL`. The third operand is always a list, which may be empty. Of course, the components may be accessed via the indexing (selection) operation.

For example, the statement

```
A := array( symmetric, 1..2,1..2, [[x,y],[y,z]] );
```

assigns to the name `A` a symmetric, 2-by-2 array whose first row contains the values `x`, `y` and whose second row contains the values `y`, `z`. The following statement would have the same effect:

```
A := array( symmetric, 1..2,1..2, [(1,1)=x, (1,2)=y, (2,2)=z] );
```

In either case,

```
op(1, eval(A));          ⟶          symmetric
op(2, eval(A));          ⟶          1 .. 2, 1 .. 2
op(3, eval(A));          ⟶          [(1,1)=x, (1,2)=y, (2,2)=z]
A[2,1];                  ⟶          y .
```

Note the use of the construct `eval(A)` here. Normally in Maple a name evaluates to its value, but if the value is a table (array) or a procedure then the result of normal evaluation is the name itself. Hence in the above example, we must specify an extra evaluation via `eval(A)` in order to extract operands from the actual array structure rather than the name `A`.

For a non-array table structure, there are only two operands which can be extracted via the `op` function: the indexing function, which may be `NULL`, and the list of components.

Maple uses a sparse representation for tables and arrays, applying a hashing technique for the internal representation. Thus there is no memory space reserved for components which have not been assigned. (See Chapter 6 for a detailed discussion of arrays and tables.)

4.1.14   Procedure Definition

A procedure definition in Maple is a valid expression and its type is named `procedure`. The external representation of a procedure definition is

$$\textbf{proc } (\ \langle \text{nameseq} \rangle \ ) \ \textbf{local } \langle \text{nameseq} \rangle; \ \textbf{options } \langle \text{nameseq} \rangle; \ \langle \text{statseq} \rangle \ \textbf{end}$$

The internal data structure represents each $\langle \text{nameseq} \rangle$ in the order shown above, followed by a (possibly null) table of remembered values, followed by the statement sequence $\langle \text{statseq} \rangle$. Since $\langle \text{statseq} \rangle$ is not a valid expression in Maple, this part of the data structure is not retrievable by the **op** function. When the **op** function is applied to this data structure, there are four operands of the structure which can be accessed: the first operand is the $\langle \text{nameseq} \rangle$ of formal parameters, the second is the $\langle \text{nameseq} \rangle$ of local variables, the third is the $\langle \text{nameseq} \rangle$ of option names, and the fourth is the **remember** table for the procedure. Any of these operands may be NULL. For example, in the procedure expression

```
proc() end
```

all four operands are NULL. As another example, consider

```
f := proc(a,b)
  local c;
  option remember;
  c := a/b;
  if type(c, integer) then c else RETURN('procname(args)') fi
end;
```

For the above procedure

op(1, eval(f));	$\longrightarrow$	a,b
op(2, eval(f));	$\longrightarrow$	c
op(3, eval(f));	$\longrightarrow$	remember

Furthermore, if the invocations `f(2,4)` and `f(4,2)` have been executed, then

op(4, eval(f));	$\longrightarrow$	table([(2,4)=f(2,4), (4,2)=2])

Note the use of the construct `eval(f)` here. Normally in Maple a name evaluates to its value, but if the value is a procedure or a table then the result of normal evaluation is the name itself. Hence in the above example, we must specify an extra evaluation via `eval(f)` in order to extract operands from the actual procedure definition rather than the name `f`. (See Chapter 7 for a detailed discussion of procedures.)

4.1.15   Unevaluated Function Invocation

One form of function invocation is

$$\langle \text{name} \rangle \ (\ \langle \text{expression sequence} \rangle \ )$$

and if $\langle \text{name} \rangle$ is undefined then the result is an unevaluated function invocation, called type **function**. Typical examples of the type **function** are `sin(x)`, `exp(x^2)`, and `g(a,b)`, assuming that `g` has not been defined and that `sin` and `exp` have their standard definitions in Maple. When the **op** function is applied to this data type, operand 0 is defined to be the name of the function and the remaining operands are the elements of the $\langle \text{expression sequence} \rangle$. For example,

```
op(0, g(a,b));              ⟶        g
op(1, g(a,b));              ⟶        a
op(2, g(a,b));              ⟶        b
op(g(a,b));                 ⟶        a,b .
```

The special type `` `!` `` is also known to the **type** function. This corresponds to the unevaluated function named **factorial** but since the special operator !is usually used for factorials, the special type name has been provided. For example, if **n** is undefined then

```
type(n!, `!`);              ⟶        true
type(n!, function);         ⟶        true
op(0, n!);                  ⟶        factorial
factorial(n);               ⟶        n! .
```

It should be noted that the expressions formed using many of Maple's operators are functions, and therefore they get represented as the data type **function** in cases where they remain unevaluated.

This is the case for the operators: **&***, **&-operators**, **$**, **@**, **@@**, **union, intersect, minus,** and **mod**. For example,

```
A &* B;                     ⟶        A &* B
type(", function);          ⟶        true
op(0, "");                  ⟶        &*
$ 1..n;                     ⟶        $(1..n)
type(", function);          ⟶        true
s union t;                  ⟶        s union t
evalb(`union`(s,t) = ");    ⟶        true .
```

4.1.16   Unevaluated Concatenation

An expression which consists of an unevaluated concatenation is said to be of type `` `.` ``. Normally the concatenation operator is evaluated to form a name, but an example of an expression of type `` `.` `` would be **a.(i^2)**, where **i** has no value. This data type has two operands which may be extracted via the **op** function: the left-hand expression and the right-hand expression.

4.1.17   Unevaluated Expression

There is a data structure for unevaluated expressions, called type **uneval**, corresponding to expressions which have been quoted (using single quotes) to prevent evaluation. For example, the expression resulting from the evaluation of

```
''x + y''
```

is the expression **'x + y'** which is of type **uneval**. This data type has only one operand that may be extracted by the **op** function: the expression inside the (outermost) pair of single quotes.

4.2   Map, Subs, and Subsop

In addition to the **type**, **nops** and **op** functions mentioned at the beginning of this chapter, the **map**, **subs** and **subsop** functions are also important functions for manipulating Maple expressions. We treat them here since they are used in the examples in later chapters.

4.2.1   Map

The **map** function replaces each operand (as defined by the **op** function) of an expression with the result of applying a function to it.[1] For example, given a list of integers, create a list of their absolute values and of their squares.

```
> L := [ -1, 2, -3, -4, 5 ];
```

$$L := [-1, 2, -3, -4, 5]$$

```
> map( abs, L );
```

$$[1, 2, 3, 4, 5]$$

```
> map( x -> x^2, L );
```

$$[1, 4, 9, 16, 25]$$

The general syntax of the **map** function is

$$\texttt{map}(\langle f \rangle, \ \langle expr \rangle, \ \langle x \rangle_1, \langle x \rangle_2, \ldots, \langle x \rangle_n);$$

where $\langle f \rangle$ is a function, $\langle expr \rangle$ is an expression, and $\langle x \rangle_1, \langle x \rangle_2, \ldots, \langle x \rangle_n$ are optional parameters. This means replace each operand of $\langle expr \rangle$ by the result of applying the function $\langle f \rangle$ to it passing any remaining arguments $\langle x \rangle_1, \langle x \rangle_2, \ldots, \langle x \rangle_n$ as additional arguments. Thus the following function calls are made

$$\langle f \rangle (\texttt{op}(1, \langle expr \rangle), \ \langle x \rangle_1, \langle x \rangle_2, \ldots, \langle x \rangle_n);$$
$$\langle f \rangle (\texttt{op}(2, \langle expr \rangle), \ \langle x \rangle_1, \langle x \rangle_2, \ldots, \langle x \rangle_n);$$
$$\ldots$$
$$\langle f \rangle (\texttt{op}(\texttt{nops}(\langle expr \rangle, \langle expr \rangle)), \ \langle x \rangle_1, \langle x \rangle_2, \ldots, \langle x \rangle_n);$$

For example

---

[1]Exception: for a table or array, the function is applied to the entries of the table or array, and not the operands or indices.

```
> L := [ seq(x^i, i=0..5) ];
```

$$L := [1, x, x^2, x^3, x^4, x^5]$$

```
> map( diff, L, x );
```

$$[0, 1, 2\,x, 3\,x^2, 4\,x^3, 5\,x^4]$$

```
> map( integrate, L, x );
```

$$[x, 1/2\,x^2, 1/3\,x^3, 1/4\,x^4, 1/5\,x^5, 1/6\,x^6]$$

Note that the result type may not necessarily be the same as the input type for algebraic types because of simplification. Consider the following examples

```
> a := 2-3*x+I;
```

$$a := 2 - 3\,x + I$$

```
> map(x -> x^2, a);
```

$$3 + 9\,x^2$$

```
> map(diff, a, x);
```

$$-3$$

Note that the symbol I is the complex unit $\sqrt{-1}$ in Maple.

4.2.2 Subs

The **subs** function is for doing *substitution*. That is, for replacing subexpressions in an expression with a new value. For example

```
> a := x^3+3*x+1;
```

$$a := x^3 + 3\,x + 1$$

```
> subs(x=y,a);
```

$$y^3 + 3\,y + 1$$

```
> subs(x=2,a);
```

The syntax of the **subs** function is

$$\text{subs(} \langle s \rangle \text{, } \langle expr \rangle \text{);}$$

where $\langle s \rangle$ is either an equation or a list or set of equations. The expression $\langle expr \rangle$ is traversed and each subexpression (i.e. each operand) in $\langle expr \rangle$ is compared with the left hand side(s) of $\langle s \rangle$. If equal, it is replaced with the right hand side(s) of $\langle s \rangle$. If $\langle s \rangle$ is a list or set of equations, then a simultaneous substitution is made. Compare

```
> f := x*y^2;
```

$$f := x \, y^2$$

```
> subs( x=y, subs( y=x, f ) );       # sequential substitution
```

$$y^3$$

```
> subs( {x=y, y=x}, f );             # simultaneous substitution
```

$$y^2 \, x$$

The general syntax of the subs function is

$$\text{subs(} \langle s \rangle_1 \text{, } \langle s \rangle_2 \text{, } \ldots \text{, } \langle s \rangle_n \text{, } \langle expr \rangle \text{);}$$

where the inputs $\langle s \rangle_1$, $\langle s \rangle_2$, ..., $\langle s \rangle_n$ are equations or lists or sets of equations. This is equivalent to the following sequence of substitutions.

$$\text{subs(} \langle s \rangle_n \text{, } \ldots \text{, subs(} \langle s \rangle_2 \text{, subs(} \langle s \rangle_1 \text{, expr)));}$$

Note, substitution only compares operands in the expression tree of **expr** with the left hand side of an equation. This is known as *syntactic* substitution. It means for example, **subs(x^2=y, x^3);** does not return **y*x** . Also, since **x+y** is not an operand of **x+y+z**, **subs(x+y=1, x+y+z);** returns **x+y+z** not **1+z** .

Note also that the result of substitution is not evaluated. It is only simplified. This is the normal way execution of functions works in Maple. That is, the arguments are evaluated, then the substitution is made, and the result is simplified. The result is not evaluated. Sometimes however, one wants an additional evaluation after a substitution. In such cases, it is necessary to use the **eval** function to force that evaluation. Compare

```
> subs( x=0, sin(x)+x^2 );
```

$$\sin(0)$$

```
> eval( subs( x=0, sin(x)+x^2 ) );
```

$$0$$

```
> subs( n=6, k=3, binomial(n,k) );
```

$$\text{binomial}(6, 3)$$

```
> eval( subs( n=6, k=3, binomial(n,k) ) );
```

$$20$$

4.2.3 Subsop

The **subsop** function replaces an operand of an expression with a value. For example

```
> f := [x,y,z];
```

$$f := [x, y, z]$$

```
> op(3,f);
```

$$z$$

```
> subsop(3=w,f);
```

$$[x, y, w]$$

```
> subsop(3=NULL,f);
```

$$[x, y]$$

The general syntax for **subsop** function is

$$\text{subsop}(\ \langle i \rangle_1 = \langle v \rangle_1, \ \langle i \rangle_2 = \langle v \rangle_2, \ \ldots, \ \langle i \rangle_n = \langle v \rangle_n, \ \textbf{expr} \ );$$

where the inputs $\langle i \rangle_1$, $\langle i \rangle_2$, ..., $\langle i \rangle_n$, must be integers, and the replacement values $\langle v \rangle_1$, $\langle v \rangle_2$, ..., $\langle v \rangle_n$ can be any expressions. This means replace (simultaneously) operand $\langle i \rangle_1$ of **expr** with $\langle v \rangle_1$, operand $\langle i \rangle_2$ of **expr** by $\langle v \rangle_2$, etc. Note that as is the case for the **subs** function, substitution is followed by simplification, but not by evaluation. Compare the following examples

```
> f := [x,y,z];
```

$$f := [x, y, z]$$

```
> subsop(1=y,2=z,3=x,f);
```

$$[y, z, x]$$

```
> f := {x,y,z};
```

$$f := \{x, z, y\}$$

```
> subsop(1=y,2=z,3=x,f);
```

$$\{x, z, y\}$$

```
> f := binomial(n,k);
```

$$f := binomial(n, k)$$

```
> subsop(1=6,2=3,f);
```

$$binomial(6, 3)$$

```
> eval(subsop(1=6,2=3,f));
```

$$20$$

5
Type Testing

Maple, as a programming language, has several attributes common to object oriented languages. In particular variables and procedure parameters can hold any valid Maple object. In principle Maple has no declarations, it is up to the functions and up to the user to enforce typing restrictions. For example

```
a:=<anything>;
f(<anything>);
subs(x=1,<anything>);
```

are examples of truly polymorphic operations. The expression `<anything>` stands for any valid Maple structure. The operations indicated will be performed with any type of operand.

In this situation, it is imperative to have a mechanism for testing the type of an expression. The function **type** is designed to do type testing by taking an expression and a type and returning **true** or **false** accordingly. For example, if the function **f** is expecting two arguments, a name and an integer, this can be checked with the following statements:

```
f:= proc( x, n )
    if nargs=2 and type(x,name) and type(n,integer) then
        <do computation>
    else
        ERROR(`incorrect arguments`)
    fi;
    . . .
end;
```

Types can be simple, or atomic. They can be predefined within Maple or defined by the user. In addition structured types can be composed from simple types or other structured types. Structured types can also be defined recursively. The remaining part of this chapter discusses the **type** function and its type argument in more detail.

Maple provides two other tools for type testing, **whattype** and **hastype**. The function **whattype** returns the top level type of an expression.

```
whattype(1)          ⟶   integer
whattype(3/2)        ⟶   fraction
whattype(x)          ⟶   string
whattype(a+b*c)      ⟶   `+`
```

Notice that **type** has knowledge about a hierarchy of types, whereas **whattype** returns the simplest surface type encountered. For example

```
type(1,numeric)      ⟶    true
type(3/2,numeric)    ⟶    true
type(1.0,numeric)    ⟶    true
```

or

```
type(1,polynom)      ⟶    true
type(1+x,polynom)    ⟶    true
type(x^2,polynom)    ⟶    true
```

Consequently when type testing in programs it is much easier and more effective to use **type** than to use an equivalent **whattype** call.

The function **hastype** returns **true** if any subexpression is an expression of the given type. For example

```
hastype(x^2+2*y,name)   ⟶    true
hastype(x^2+2*y,float)  ⟶    false
```

5.1   Definition of a Type in Maple

A type is defined by a domain or subdomain of computation. We will say that x is of type y when x belongs to the domain of computation named y. Some types are defined by classical domains such as **integer**, **rational**, or **polynom** (Maple's type name for polynomials). Other types are closely linked to the internal data structures such as **list**, **set**, or **equation**. In Maple, a type is any expression which is recognized by the type function and causes it to return true for some set of expressions and false otherwise. The expression

```
type(a,t);
```

evaluates to **true** if **a** is of type **t** and to **false** otherwise. Formally the term *type t* will be used to refer to the set of all expressions which type-test to true with argument t. The expressions to test for types, or *type-expressions,* can be constructed, used for testing, passed as parameters, split into their components, merged, or simplified. As such we have a set of operations (an algebra) over type-expressions.

A type can be a *simple type* or a *structured type*. A *simple type* is one which is identified by a single name. For example: **integer**, **relation**, **list**. A *structured type* is a Maple expression which allows, in general, a more thorough type analysis. For example: **list(integer)**, **polynom(rational)**.

5.2   Simple Types

A simple type can be:

(a) a *system type* when the type is a name and it is defined in the kernel of the system. System types usually correspond to primitive algebraic or data structure concepts. For example

```
type([a,b,c],list)    ⟶    true
type(10,float)        ⟶    false
type(a=b,relation)    ⟶    true
```

(b) a *procedural type* when the type is a name, such as **xxx**, and there is a procedure `'type/xxx'` that will perform the type analysis of the argument and will return true or false accordingly. This procedure may be available from the global Maple environment or from the library (when in the library it is automatically loaded into the environment). This is one of the mechanisms to define new types in Maple. For example, **monomial**, **algnum,** and **odd** are presently implemented as external Maple functions. This allows us to see the definition of the type as follows

```
> interface(verboseproc=2);
> print(`type/odd`);

  proc(n)
  options `Copyright 1990 by the University of Waterloo`;
    type(n,integer) and (irem(n,2) = 1)
  end
```

(c) an *assigned type* when the type is a name, such as **xxx** , and the global name `type/xxx` is assigned a type expression. The type evaluation proceeds as if the type checking were done with the expression assigned to `type/xxx` . For example, to check the correctness of the arguments to the function int we use

```
`type/intargs` := [algebraic, {name, name=algebraic..algebraic}];
    . . .
if not type([args],'intargs') then ERROR(...) .
```

(d) a *constant* when it is a numeric value. This type will match an expression which is identical to the type.

```
type(1,1)         ⟶    true
type(1.99,2.00)   ⟶    false
```

5.3   Structured Types

Structured types are Maple expressions which allow a more detailed type testing. These expressions are built using any of the Maple operators, constants, and the names of simple types. There are two classes of structured types in Maple:

(a) An expression which mimics the expression to be tested. For example, X Op Y tests for the binary operator Op with a first argument of type X and a second argument of type Y.

In this case the type-expression is a composite expression and the testing of `type(a, t)` proceeds through the following steps, referring to the function invocation

(i) check that the top level operation of `a` and `t` are the same;

(ii) check that the number of subexpressions of `a` and `t` are the same (so `nops(a)=nops(t)`);

(iii) check that the ith subexpression of `a` is of the type defined by the ith subexpression of `t` .

For example, the structured type `[algebraic, name=integer..integer]` matches `[x + y, x = 1..100]` .

This is a very powerful and mnemonic technique to define types. Table 5.1 makes these definitions more precise.

The type **anything** , which matches any Maple structure, becomes a useful type when used with type expressions. For example,

```
{ name = anything, anything = name } .
```

(b) A *hierarchical type* , namely, a type of the form `x(y)` which matches all the expressions of type `x` for which all of their subexpressions are of type `y` . For example, `` `+`(name) `` defines the type of all expressions which are sums and whose components are all names.

```
type( x+y, `+`(name) )     ⟶    true
type( x+2*y, `+`(name) )   ⟶    false
```

In general, the test `type(expr, x(y))` is equivalent to

```
if not type(expr, x) then RETURN(false)
else for v in expr do
    if not type(v, y) then RETURN(false) fi od;
fi;
RETURN( true )  .
```

For example, `list(numeric)` matches `[1, 2/3, 4.5]` and `[]` but does not match `{1, 2}` nor `[{}, 1]`.

The functional notation was chosen because its common oral reading conveys the desired interpretation. That is to say, *X(Y)* is read as "*X* of *Y*" which is a good semantic definition of a hierarchical type.

Some system-defined types give a special interpretation to the meaning of *subexpression*. For example, for the type **polynom** , the subexpressions are understood to be the coefficients of the polynomial. So `polynom(integer)` type-matches polynomials with integer coefficients.

If the type being defined does not have subexpressions then it cannot be used in this form. For example, it is incorrect to use `integer(name)` since an integer is a primitive object and has no subexpressions. Similarly it is useless, although correct, to use `rational(integer)` because the two components of a rational are always integers.

Note that the use of structured types, are not only more succint, but also more efficient. Compare the next two equivalent tests to appreciate this difference.

Syntactic definition	Matches
type ::= { *type*, ... }	alternation; any of the types
	[*type*, ...]	a list of the given types
	numeric	match a numerical constant exactly
	string	a system, procedural, or assigned type
	type = *type*	an equation of the corresponding types
	type <> *type*	an inequality of the corresponding types
	type < *type*	a relation of the corresponding types
	type <= *type*	a relation of the corresponding types
	type > *type*	a relation of the corresponding types
	type >= *type*	a relation of the corresponding types
	type .. *type*	a range of the corresponding types
	type **and** *type*	an **and** of the corresponding types
	type **or** *type*	an **or** of the corresponding types
	not *type*	a **not** of the corresponding type
	type &+ *type* ...	a sum of the corresponding types
	type &* *type* ...	a product of the corresponding types
	type ^ *type*	a power of the corresponding types
	type . *type*	a concatenation of the corresponding types
	' *type* '	an unevaluated expression of the given type
	name [*type*, ...]	an indexed reference of the given types
	typename(*subtype*)	an expression of type *typename*, of which the subexpressions are of type *subtype* (see (b)).
	identical(*expr*)	an expression identical to *expr*
	specfunc(*type* , *foo*)	the function *foo* with *type* arguments
	anyfunc (*type*, ...)	any function of the given types
	foo (*type*, ...)	the function *foo* of the given types

TABLE 5.1. Structured Types

```
type( expr, name=numeric..numeric )
type( expr, equation) and type(op(1,expr),name) and type(op(2,expr),range)
and type(op(1,op(2,expr)),numeric) and type(op(2,op(2,expr)),numeric)
```

Due to the syntax used for hierarchical type testing, the two classes of structured types have an area of overlap for unevaluated function testing. This overlap is resolved in the following way.

(a) A type expression `a(b)` , when `a` is a valid type, is treated as a hierarchical type expression. For example

```
type(f(1,2,3), function(integer))
```
$\longrightarrow$ true

(b) A type expression `a(b)` , when `a` is not a valid type, is considered to be a type-mimicking expression. For example,

```
type(exp(1+x), exp(algebraic))
```
$\longrightarrow$ true

(c) A specific function with one or many arguments can be tested with the special form: `specfunc(type, functname)` . For example,

```
type(GAMMA(50, Pi), specfunc(constant, GAMMA))
```
$\longrightarrow$ true

(d) An arbitrary function with specific arguments can be tested with the special form: `anyfunc(type1, type2, ...)` . For example,

```
type(f(1, 2/3), anyfunc(integer, fraction))
```
$\longrightarrow$ true

In the terminology of formal language theory, if we consider the system and procedural types as special (terminal) symbols then there is a one-to-one correspondence between context-free grammars and type definitions. In other words, in terms of system and procedural types, we can give a type definition which will test for any context-free language of type expressions.

5.4   Surface and Nested Types

The type checks which require information about the top level of the expression tree alone will be called *surface types*. The types which check a complete expression tree (probably recursively) will be called *nested types*. Most of the system types are surface types since they are coded in the top node of the expression tree. For example, with

```
type( { .. }, set );
type( [a] + [b,c], algebraic );
```

both return **true** regardless of the types of the components of the set in the first case, and regardless of the types of the terms of the sum in the second case. Both of these types are surface types.

The type **constant** , on the other hand, will completely scan an expression to determine whether or not it is composed of any non-constant parts. Hence it is a nested type. Any assigned type corresponding to a nested type requires a (direct or indirect) recursive definition. It is because of this need for recursion that an assigned type requires assignment to a name prefixed by `` `type/` ``, which is different from the type name itself. Otherwise these objects would have a recursive definition and they would be very difficult to handle without spawning an infinite evaluation. For example, one

$$
\begin{array}{rcl}
`**` & \equiv & `^` \\
`..` & \equiv & \text{range} \\
`=` & \equiv & \text{equation}
\end{array}
$$

TABLE 5.2. Equivalent Types

```
                                   {integer, fraction}  ≡  rational
                                    {rational, float}   ≡  numeric
                                    {string, indexed}   ≡  name
  {numeric,name,`+`,`*`,`^`,series,function,`.`,uneval} ≡  algebraic
                            {`=`, `<>`, `<`, `<=`}      ≡  relation
```

TABLE 5.3. Disjoint Identities

could define

```
`type/Boolean` := { name, function, relation, not Boolean,
`and`(Boolean), `or`(Boolean) };
type(a or not c=d , Boolean)  ⟶  true
```

5.5 Simplification of Types

The ability to define types as valid Maple expressions gives us the possibility of manipulating these types. For example, we can build new types, pick up parts of types, and pass types as parameters. A useful manipulation is the minimization and simplification of types.

The basic types involve some inclusion/complement relationships which form part of the axioms of the type algebra. Tables 5.2 through 5.4 illustrate the basic relationships and help to explain the overall type hierarchy.

From these tables we can derive the following simplification rules.

(a) Canonical equivalencing. This is the case of transforming a set of types into an equivalent type by using any of the transformations from the first two tables. For example,

```
integer, fraction, rational, float  ⟶  numeric .
```

(b) Redundant inclusions. In this case, we use the closure of the third table and the expression types to simplify redundant inclusions. For example,

```
radical, algebraic       ⟶  algebraic
name=integer, relation   ⟶  relation .
```

(c) All of the simplifications which can be applied to context-free grammars can be applied to types. Unfortunately there is no minimal or normal-form expression for a context-free language;

$$
\begin{array}{rcl}
\texttt{radical} & \subset & \texttt{`^`} \\
\texttt{polynom} & \subset & \texttt{ratpoly} \\
\texttt{ratpoly} & \subset & \texttt{algebraic} \\
\texttt{`!`} & \subset & \texttt{function} \\
\texttt{\&+} & \subset & \texttt{`+`} \\
\texttt{\&*} & \subset & \texttt{`*`} \\
\texttt{numeric} & \subset & \texttt{constant} \\
\texttt{constant} & \subset & \texttt{algebraic} \\
\texttt{\{relation,`and`,`not`,`or`\}} & \subset & \texttt{boolean} \\
\texttt{set(..)} & \subset & \texttt{set} \\
\texttt{list(..)} & \subset & \texttt{list} \\
\texttt{array} & \subset & \texttt{table} \\
\texttt{\{algebraic,boolean,set,list,table\}} & \subset & \texttt{anything}
\end{array}
$$

TABLE 5.4. Proper Subsets (non-derivable)

consequently there will be none for types.

(d) Ad-hoc type simplifications such as the examples shown in Table 5.5 .

5.6 Parameter Type Testing

By far the most common need for type testing is as the first step of function invocation. Unless a variable number of arguments are required, or unless the types of some arguments depend on other arguments, one type expression is enough to do complete type checking. Recall the first example of the function f which required two arguments, the first a name and the second an integer. The entire test can be done with

```
if not type([args],[name,integer]) then ERROR(...) fi;
```
Notice that by enclosing all the arguments in a list, [args], we also achieve the argument count, nargs=2, together with type testing. This type of testing is recommended due to its readability and efficiency.

5.7 Undesirable Simplifications and Evaluations of Types

A type expression is a general Maple expression which is subject to all of the standard evaluations and simplifications. The first problem is raised by some simplifications which are done automatically

top recursion removal:		
`` `type/X` `` := { <non-X-type>, X };	$\longrightarrow$	`` `type/X` `` := <non-X-type>
flattening of alternation:		
{a,{b,c},{b,c,d}, ... }	$\longrightarrow$	{a,b,c,d, ... }
{a}	$\longrightarrow$	a
factoring common operations:		
{ f(a), f(b), f(c) }	$\longrightarrow$	f({a,b,c})
impossible types:		
{integer^integer}	$\longrightarrow$	{}
unnecessary generality:		
anything=anything	$\longrightarrow$	equation

TABLE 5.5. Ad-Hoc Type Simplifications

to any expression. For example, the system will automatically transform

    x and x $\longrightarrow$ x

for any **x** and hence

    Boolean and Boolean $\longrightarrow$ Boolean

Consequently the natural definition of a nested type **Boolean** , as presented above, will not work due to automatic simplification. This problem and similar ones can be solved by using the hierarchical type notation, given that both arguments of the **and** are of the same type. For example, `` `and` ``(Boolean) will not change under automatic simplification.

    A second problem generated by simplification is the possible arbitrary order that may result for some commutative operators. For example, **integer &+ name** will match **1+x** but this expression may be simplified to **x+1** automatically by the internal simplifier, and the latter will not match the given type expression. In this case, the only solution is to test for both orderings, as in {**integer &+ name, name &+ integer**} . Products give a similar problem except that rational coefficients are always placed at the front of the expressions. Hence **integer &* name** will match both **x*3** and **3*x** since the former is always converted to the latter form.

In general, these problems are caused by pattern matching problems rather than by data structure type tests. For sums and products the Maple function **match** is probably what is desired by most users.

    Evaluation may also cause some unwanted effects. Suppose that we want to test for an unevaluated indefinite integral, an integral which could not evaluate because it does not have a closed form. An example would be

    int(exp(x^3), x) .

Using structured types we would test for this with the type **int(algebraic,name)**. Evaluation plays us an unexpected trick here. The above type expression is really a call to **int** that results in `algebraic*name` as if the names involved were variables. This problem of evaluation arises when type-testing for unevaluated functions and for some type names which are also function names, such as `matrix` or `array`. The solution to this problem is to quote the function name, which prevents function evaluation while allowing for proper argument evaluation. For example,

```
type( expr, 'int'(algebraic,name) )
type( A, 'matrix'(float) ) .
```

5.8   Type Testing Versus Pattern Matching

The functionalities of type testing and pattern matching have an area of overlap. It is important to define what is considered type testing and what is considered pattern matching. An example of the overlap is the following test for linearity in **x**.

```
type( expr, (integer &* identical(x)) &+ integer );
match( expr = a*x+b, x, 's' );
```

We define *type testing* to be a *data structure* or *form* test which is independent of the underlying equivalences derived from the algebraic properties of operators. In other words, type testing will test expressions without assuming any knowledge about simplification rules, associative laws, commutative laws, distributive laws, or other algebraic properties. It is purely a *syntactic* test. For example,

`2*x+3`	matches the type (`integer &* identical(x)) &+ integer`
`2*x`	does not match the above type
`3`	does not match the above type
`x+3`	does not match the above type

Note that pattern matching will match all of the above expressions to the pattern **a*x+b**, after making appropriate choices for the parameters **a** and **b**.

```
match(2*x+3, a*x+b, x, `r`) ⟶   true
`r`                          ⟶   true
```

6
Arrays and Tables

6.1 Overview

One of the data types in Maple is the *table* structure. The *array* structure in Maple is a specialization of the table data structure in which the indices must be integer expression sequences lying within user-specified bounds. Arrays are used similarly to those in other programming languages, while tables correspond roughly to the ones provided in Snobol or Icon.

As with other data types in Maple, tables are self-describing data objects, which may be created dynamically, passed as parameters, and so on. No declarations are needed; to make a name refer to a table, an assignment statement is used in which the right-hand side evaluates to a table object.

A `table` object consists of three parts:

- an indexing function,
- an index bound (for arrays only), and
- a collection of components.

The indexing function allows a table to have a user-defined interface. A detailed discussion of indexing functions is given in Section 6.5. If no special interface is to be used, the indexing function should have the value `NULL`, which is the default.

The only tables which have index bounds are arrays. The index bound is an expression sequence of integer ranges. The number of ranges is called the *dimension* of the array. Whenever a component of an array is referenced, the index is checked against the index bound. The ith range gives the bounds on the ith integer in the expression sequence used as the index.

The `op` function may be used to extract the operands of a table or array. For an array structure, the indexing function is available as the first operand, the index bound is the second operand, and the third operand is the collection of components specified as a list of equations. Either of the first two operands may be `NULL`, while the third operand is always a list, which may be empty. For a non-array table structure, there are only two operands which can be extracted via the `op` function: the indexing function (which may be `NULL`), and the list of components.

Components of tables may be accessed by using brackets ('[' and ']') for indexing. If the value of `T` is a table then components of `T` are referenced using the syntax

```
T [ ⟨expression sequence⟩ ].
```

For example, executing the following statements

```
T[1,2] := a;
T[2,0] := b;
V := T[1,2] + T[2,0];
```

causes V to be assigned the value a+b.

Tables may be created either (i) explicitly, by calling one of the builtin functions **array** or **table**, or (ii) implicitly, by making an assignment to an indexed name of the form

```
A [ ⟨expression sequence⟩ ]
```

when A does not have a previous value. The creation of tables is described in Section 6.2.

Expressions of type **array** and of type **table** are represented internally using the same data structure. The external representation is as a call to one of the functions **array** or **table** which would re-create the object. Specifically, the external representation for a table is:

```
table( ⟨indexing function⟩ , [ ⟨equation sequence⟩ ])
```

where each equation in the ⟨equation sequence⟩ is of the form

```
(index) = component_value   .
```

This format may be used for the input of a table structure. For output, note that the equation sequence enclosed in brackets is a representation of an internal hash table and therefore the equations will appear in an apparently arbitrary order which cannot be controlled by the user.

The external representation for an array is:

```
array( ⟨indexing function⟩ , ⟨range sequence⟩ , ⟨values⟩ ) .
```

This format may be used for the input of an array structure, where the ⟨values⟩ may be specified as a list of equations (exactly as in the case of tables mentioned above) or the ⟨values⟩ may be specified as a list of lists (a row-by-row representation of the array). For output in Maple's pretty-printing mode, one-dimensional arrays are printed in the form of a row vector, two-dimensional arrays are printed in a two-dimensional matrix notation, and for higher-dimensional arrays the output closely resembles the input notation but with the ⟨values⟩ presented as an ordered sequence of equations. In line-printing mode, the output closely resembles the input notation with the ⟨values⟩ of the array printed as an unordered ⟨equation sequence⟩ as described above for general tables.

6.2   Creating Tables

6.2.1   Explicit Table Creation

The function **array** is used to create an array explicitly. To create a table that is not an array, the function **table** is used. These functions take a number of optional parameters which specify information about the table to be created.

The most common uses of these functions are illustrated by the following examples:

```
t := table();
a := array(1..n);
b := array(1..n, 1..m);
```

Here, **t** has been assigned a new table object, **a** has been assigned a one-dimensional array with **n** components, and **b** has been assigned an **n** by **m**, two-dimensional array.

The **table** function, in general, takes two parameters: an indexing function and an initialization list. The function **array** takes an indexing function, an initialization list, and an index bound. The index bound is passed as a number of integer ranges appearing adjacently in the parameter sequence. The indexing function, initialization list and, for arrays, index bound are all optional and may appear in any order in the parameter sequence.

When a table is created using one of these functions, the following sequence of events takes place:

1. The parameters are sorted out.

2. If no indexing function is supplied, **NULL** is taken as the default.

3. If no initialization list is supplied, an empty list is taken as the default.

4. If it is an array that is being created and no index bound has been supplied, the index bound to be used is deduced from the initialization list.

5. If the initialization list is not empty, the initial values are inserted into the table.

6. The table is returned.

The indexing function must be given as either a procedure or as a name. Not specifying an indexing function is the only way to obtain **NULL**.

The deduction of index bounds for arrays and the initialization of table values are done by two procedures from the Maple library. The initializations are specified as a list of equations, as a list of values, or as a list of lists. With a list of equations, the left-hand sides are the indices (of the components to be initialized) and the right-hand sides are the values. A list of values may be used to specify the creation of a table or a one-dimensional array, in which case the indices used are

Maple Input	Result
`table( );`	`table([ ])`[1]
`table([a,b,c]);`	`table([(1)=a,(2)=b,(3)=c])`
`table([1=a0, cos(x)=a1]);`	`table([(1)=a0,(cos(x))=a1])`
`array(0..3);`	`array(0..3,[ ])`
`array([x,y,z]);`	`array(1..3,[(1)=x,(2)=y,(3)=z])`
`array([b,c,d], 0..3);`	`array(0..3,[(0)=b,(1)=c,(2)=d])`
`array([NULL=val1]);`	`array([()=val1])`
`array([(2,2)=22, (1,7)=17]);`	`array(1..2,2..7,[(2,2)=22,(1,7)=17])`
`array(sparse,[5=x,100=y]);`	`array(sparse,5..100,[(5)=x,(100)=y])`
`array([[1,2],[3,4]]);`	`array(1..2,1..2,`
	`[(1,1)=1,(1,2)=2,(2,1)=3,(2,2)=4])`

TABLE 6.1. Explicit Table Creation

consecutive integers starting at 1 (or the lower bound on the indices, if one is given, for an array). For two-dimensional (or higher-dimensional) arrays, the initializations may be specified as a list of the rows of the array, where each row is itself specified as a list (or a list of lists if there are more than two dimensions).

If no index bound is given for an array, then one is deduced from the list of initializations. This is done as follows. If the initialization list is empty, then the array is assumed to be zero-dimensional and the index bound is NULL (thus a sequence of zero integer ranges). If the initialization is given as a list of n values, then the array is taken to be one-dimensional and the index bound is the range `1..n`. If the initializations are given as a list of equations, then each range in the index bound is made as restrictive as possible while still encompassing all the indices used in the equations. Finally, if the initialization is given as a list of lists then the number of dimensions is deduced from the level of nesting of the lists and each dimension is indexed from 1.

Some examples are shown in Table 6.1 .

6.2.2 Implicit Table Creation

A table is implicitly created if an assignment is made to an indexed name of the form T[⟨expression sequence⟩] and T does not have a previous value. When T does not have a previous value, then the assignment

    T[eseq] := expr;

is exactly equivalent to the following statement sequence which uses explicit table creation:

---

[1]The indexing function is NULL and no components have been initialized.

```
    T := table();   T[eseq] := expr;
```
This rule is applied recursively if necessary.

Examples:

If `A` is a table but `A[1]` has not been assigned, then
```
    A[1][2,x] := y;
```
is equivalent to
```
    A[1] := table();   A[1][2,x] := y;
```
If `B` does not have a previous value, then
```
    B[i,k][j] := f(i,j);
```
is equivalent to
```
    B := table();   B[i,k] := table();   B[i,k][j] := f(i,j);
```


6.3   Evaluation Rules for Tables and Table Components

6.3.1   Evaluation to a Table

Evaluation of an expression in Maple is normally performed in a full, recursive evaluation mode. The result of evaluation is normally the fully evaluated expression. One exception to this rule is when the evaluated expression is a table (or array) and there is an identical exception for a procedure. (See Section 7.1 for the evaluation rules for procedures.) If evaluation of an expression yields a table structure then the result of evaluation is the *last name evaluated*.

The most common situation is the case of a name `T` which has been assigned a table (or array) structure, in which case the result of evaluating `T` will be the name `T` rather than the value of `T`. If the value of `T` is required, the construct `eval(T)` must be used. In some contexts there will be no name appearing in the evaluation chain, in which case there is no *last name evaluated*; in such a situation, the result of evaluation will be the table structure.

Examples:

`T[x] := a;`	$\longrightarrow$	`T[x] := a`
`eval(T);`	$\longrightarrow$	`table([(x)=a])`
`f(T);`	$\longrightarrow$	`f(T)`
`S := T;`	$\longrightarrow$	`S := T`
`S[x];`	$\longrightarrow$	`a`
`f(S);`	$\longrightarrow$	`f(T)`
`eval(T);`	$\longrightarrow$	`table([(x)=a])`
`f(");`	$\longrightarrow$	`f(table([(x)=a]))`

In particular, note the result of evaluating the expression f(S) above. At that point, S points to the name T which points to the table structure. Thus full evaluation of S leads to the table structure and the result of this evaluation is T because the latter is the *last name evaluated* in the chain of evaluation. Note also the final statement f("). In this case, the evaluation of the function argument leads to a table structure and there is no name in the evaluation chain, so the table structure itself is the result of the evaluation. Of course, the exact same behaviour would occur if the value of T were an array.

6.3.2   Evaluating Table Components

The semantics of referencing a table's components are defined by its indexing function. With the default indexing function, NULL, the usual notion of subscripting is used. With other indexing functions, a procedure determines how indexing is done. This more complicated case is discussed in Section 6.5. In this section, the default indexing semantics are described.

Suppose that the value of T is a table with a NULL indexing function. When T[⟨expression sequence⟩] is evaluated, the value of the entry in the table is returned, if there is one. If there is not an entry with the ⟨expression sequence⟩ as its key, then the table reference "fails" in the following sense. The value returned is an indexed name where the index is the ⟨expression sequence⟩, evaluated, and the zeroth operand is the *last name evaluated* in the evaluation chain for T.

Examples:

```
t := table();              ⟶        t := table([ ])
t[k] := ZZ:
eval(t);                   ⟶        table([(k)=ZZ])
s := t:
s[1] := XX:
eval(s);                   ⟶        table([(1)=XX,(k)=ZZ])
t[k];                      ⟶        ZZ
s[1];                      ⟶        XX
t[2];                      ⟶        t[2]
j := 2:
t[j];                      ⟶        t[2]
s[2];                      ⟶        t[2]
p := proc(a) a[2] end:
p(s);                      ⟶        t[2]
```

In the above examples, the name s evaluates to the name t, whose value is the table. That is why s[2] yields t[2] when it "fails". In the procedure call p(s), the name s evaluates to the name t which is passed in as the argument to the procedure.

Even if the last name in the evaluation chain evaluates to some other object before evaluating to the table, it is still used if a table reference "fails".

Examples:

```
a := f(table( ));              ⟶        a := f(table([ ]))
f := proc(t) print(hello); t end:
a[x];                          ⟶        a[x]
```

after printing `hello`.

An array is zero-dimensional if its index bound is the null expression sequence. A zero-dimensional array has only one component and the index for this component is NULL.

Examples:

```
t := array( );                 ⟶        t := array([ ])
t[ ] := tval:
t[ ];                          ⟶        tval
```

6.3.3   Assigning and Unassigning Table Components

If T is a table and an expression of the form T[⟨expression sequence⟩] appears on the left-hand side of an assignment statement, then an entry is assigned in the table. If there was previously no entry in the table with the evaluated ⟨expression sequence⟩ as its key, then a new entry is made. If there already is an entry, then it is updated to reflect the new value.

In many cases it is desired to assign a value to a parameter of a procedure. Table components may be assigned this way, in the same way as names. To assign a table component, an indexed name is passed. Consider the procedure

```
assignsqr := proc(a,b) a := b^2 end
```

So long as the first parameter is a valid left-hand side, the assignment will be made.

Examples:

```
t := table();
assignsqr(t[2], 4);        assigns the value 16 to t[2]
s := 's';                  unassigns s
assignsqr(s[3], 3);        assigns 9 to s[3], implicitly creating a table
```

If the component to be assigned already has a value, then it is necessary to use quotes or the **evaln** function to pass the indexed name rather than its value.

Examples:

After executing the statements

```
t := table();
for i to 5 do assignsqr(t[i], i) od;
```

assigning new values to the components of **t** may be achieved by

```
for i to 5 do assignsqr(evaln(t[i]), 1/i) od;
```

When the subscript need not be evaluated, quotes may be used:

```
assignsqr('t[1]', x);
```

To make a name stand for itself in Maple, a statement is executed to "unassign" it. Exactly the same thing is done with the components of a table — to remove an entry from a table, it is "unassigned". This may be done either by quoting the right-hand side or by using the **evaln** function.

Examples:

```
a := array([x, y, z]):
a[1];                           ⟶    x
a[1] := 'a[1]':
a[1];                           ⟶    a[1];
lprint(eval(a));                ⟶    array(1..3,[(2)=y,(3)=z])
i := 3:
a[i] := evaln(a[i]):
lprint(eval(a));                ⟶    array(1..3,[(2)=y])
```

To determine whether a table component is assigned, the **assigned** function is used. A component **T[x]** is deemed to be assigned if the evaluation of **T[x]** would yield a different result than **evaln(T[x])**. Thus a definition for **assigned(T[x])** is:

```
evalb( T[x] <> evaln(T[x]) ) .
```

Examples:

```
a := array(1..3, [e,f]):
a[2];                           ⟶    f
assigned(a[2]);                 ⟶    true
assigned(a[3]);                 ⟶    false
a := table(sparse):
a[v];                           ⟶    0
evaln(a[v]);                    ⟶    a[v]
assigned(a[v]);                 ⟶    true
```

6.3.4 Table Operations

Consider storing a table called **Deriv** of some of the derivatives of the elementary functions. Our table might look like this.

```
> Deriv := table( [exp=exp, ln=(f->1/f), sin=cos, cos=-sin] );
```

```
Deriv := table([
                 cos = - sin
                 exp = exp
                 sin = cos
                 ln = f -> 1/f
             ])
```

We have seen how entries in a table are accessed and how new entries can be inserted or assigned, and how they can be removed or unassigned. For example

```
> Deriv[sin];
```

$$- \cos$$

```
> Deriv[tan] := 1+tan^2:
> Deriv[exp] := 'Deriv[exp]':
> print(Deriv);
```

```
                        table([
                            cos = - sin
                            sin = cos
                            ln = f -> 1/f
                                            2
                            tan = 1 + tan
                        ])
```

As was the case for lists, sets, and other Maple structures, the *components* or *operands* of tables can be extracted using the **op** function as follows

```
> nops(eval(Deriv));
```

$$2$$

```
> op(1,eval(Deriv));
> op(2,eval(Deriv));
```

$$[\cos = - \sin, \ \sin = \cos, \ \ln = f \rightarrow 1/f, \ \tan = 1 + \tan^2]$$

A table has 2 operands. The first is the indexing function[2]. In this example, the result is the empty sequence NULL. The second operand is a list of index-entry pairs returned as equations where the left hand side is a table index, or key, the right hand side its corresponding entry, or value. Arrays have 3 operands. The first operand is the indexing function, the second is a sequence of the dimensions, and the third is a list of the index-entry pairs. For example

```
> A := array(1..2,1..2,[[1,u],[1,v]]);
```

$$A := \begin{bmatrix} 1 & u \\ 1 & v \end{bmatrix}$$

```
> nops(eval(A));
```

_____

[2]The meaning of indexing functions is explained later in this chapter.

$$3$$

```
> op(1,eval(A));
> op(2,eval(A));
```

$$1 \; .. \; 2, \; 1 \; .. \; 2$$

```
> op(3,eval(A));
```

$$[(2, 1) = 1, (2, 2) = v, (1, 1) = 1, (1, 2) = u]$$

Two other useful functions for accessing components of tables are the **indices** and **entries** functions. The **indices** function returns a sequence of the indices of a table. Likewise the **entries** function returns a sequence of the entries of a table. For example

```
> indices(Deriv);
```

$$[cos], \; [sin], \; [ln], \; [tan]$$

```
> entries(Deriv);
```

$$[- \; sin], \; [cos], \; [f \; \rightarrow \; 1/f], \; [1 + tan^2]$$

Notice that each index and entry is returned in a list. This is because each index and entry of a table can itself be a sequence. Consider the following examples

```
> indices(A);
```

$$[2, \; 1], \; [2, \; 2], \; [1, \; 1], \; [1, \; 2]$$

```
> entries(A);
```

$$[1], \; [v], \; [1], \; [u]$$

```
> Colours := table( [red=(rot,rouge),blue=(blau,bleu),yellow=(gelb,jaune)] );
```

```
                    Colours := table([
                            blue = (blau, bleu)
                            red = (rot, rouge)
                            yellow = (gelb, jaune)
                    ])
```

```
> indices(Colours);
```

$$[blue], \; [red], \; [yellow]$$

```
> entries(Colours);
```

$$[blau, \; bleu], \; [rot, \; rouge], \; [gelb, \; jaune]$$

Note that the ordering in which the indices (and entries) are extracted cannot be controlled by the user because tables are implemented using hash tables. See Section 9.3 for further details about the implementation. However, there is a one to one correspondence between the order of the indices and the entries.

6.4 Tables as Objects

6.4.1 Copying Tables

In Maple, only objects of type `table` may be altered after having been created. This is because it is only with tables that it is valid to make an assignment to a *part* of the object.

The fact that a table object may be altered after creation means that if two names evaluate to the same table, then an assignment to a component of one affects the other as well. To illustrate, if the following statements are executed:

```
a := array([t,x,y,z]);
b := a;
a[1] := 9;
```

then `b[1]` will evaluate to 9, not `t`. Even if we use the statement

```
b := eval(a)
```

in place of the statement

```
b := a
```

in the above example, we get the same behaviour. In the former case, the name `b` points to the name `a` which points to the array structure; in the latter case, `b` points directly to the array structure as does `a`. In both cases there is only one array object in existence, and after the assignment `a[1] := 9` (or equivalently `b[1] := 9`) then the value of the first component in the array object will be 9.

For this reason the `copy` function is provided (see the Maple Library). It may be used to create a copy of a table upon which operations may be performed without altering the original.

For example, if a procedure makes assignments to components of a table passed as a parameter, then it may be necessary to pass a copy. Suppose that `decomp` has been assigned a procedure that does an in-place LU matrix decomposition, and takes an array as its only parameter. If it is desired to find the LU decomposition of the matrix given by `a` while retaining `a` for further use, then the following statements may be used:

```
b := copy(a);
decomp(b);
```

6.4.2 Mapping Functions Onto Tables

The `map` function can be used to apply a function to the entries of a table, creating a new table. The indices of the table remain the same. For example, given an array of polynomials compute an array of their derivatives

```
> A := array( 1..4, [x,2*x^2-1,4*x^3-3*x,8*x^4-8*x**2+1] );
```

$$A := [\, x, \; 2\,x^2 - 1, \; 4\,x^3 - 3\,x, \; 8\,x^4 - 8\,x^2 + 1\,]$$

```
> map(diff,A,x);
```

$$[1, 4 x, 12 x^2 - 3, 32 x^3 - 16 x]$$

Note, this is equivalent to the following

```
> B := array( 1..4 ):
> for i in [indices(A)] do B[op(i)] := diff(A[op(i)],x) od:
> print(B);
```

$$[1, 4 x, 12 x^2 - 3, 32 x^3 - 16 x]$$

6.4.3   Tables Local to a Procedure

A variable local to a procedure may be assigned a table, just as it may be assigned an object of any other type. A table object which is created and assigned to a local variable may be returned as the value of the procedure or passed out through one of the parameters, in a manner similar to any other expression. The main distinction with tables is that a name whose value is a table evaluates to its own name, not to the object. Therefore, it is necessary to use the **eval** function if the table object is to be obtained. In particular, if a local variable T has been assigned a table and this table is to be returned from the procedure then one must be careful to return the value **eval(T)** rather than simply T. (Returning a local variable name out of a procedure is very unwise.)

Examples:

```
# Put the coefficients of a polynomial in a table indexed by the terms.
getcoeffs := proc(poly, var)
    local Cs, c, terms, i;
    if not type(poly, polynom(anything, var)) then
        ERROR(`expecting a polynomial`)
    fi;
    Cs := table();
    c := coeffs(collect(poly,var), var, 'terms');
    for i to nops([c]) do
        Cs[terms[i]] := c[i]
    od;
    eval(Cs)
end;
Cs := table([this, that]);        ⟶      Cs := table([(1)=this,(2)=that])
getcoeffs(3*x^67 + y, x);         ⟶      table([(1)=y,(x^67)=3])
eval(Cs);                         ⟶      table([(1)=this,(2)=that])
```

6.4.4   Tables as Parameters

A table may be passed as a parameter into or out of a procedure. Components added to the table or removed from the table while the procedure is executing affect the globally visible table, since it is the same object.

Passing an unnamed table object as a parameter may lead to awkward results if components which do not have values are used. If the procedure makes a component reference that "fails" and assigns it to another component of the same table, then doing the assignment creates a self-referential data structure. (Just as x := y; y := 'x^2' does.) This would lead to an infinite evaluation recursion the next time the component was evaluated. It is recommended to always pass, as a parameter, the name of a table (such as A) rather than the table structure (eval(A) if the value of A is a table structure).

6.4.5   Automatic Loading of Tables

It is possible to define large tables that get loaded only when a component is referenced. This is done in the same way that procedures can be made to be read in only when used.

To cause a table to be loaded automatically, it is assigned an unevaluated call to **readlib**. For example, to make T be loaded when it is used, one can make the assignment

```
T := 'readlib('T', filename)';
```
where **filename** is the name of the file in which the table has been saved. (See Section 7.10 for a discussion of the **readlib** function.)

Suppose a user enters Maple and executes the following statements:

```
Linverse := table();
Linverse[1/s^n] := t^(n-1)/(n-1)!;
Linverse[1/(s^2 + a^2)] := sin(a*t)/a;
save `/u/jqpublic/laplace.m`;
quit
```
If in a subsequent Maple session the assignment

```
Linverse := 'readlib('Linverse', `/u/jqpublic/laplace.m`)';
```
has been made, then evaluating

```
Linverse[1/s^n]
```
causes **readlib** to be executed and the table is read in. The selection operation then evaluates to

```
t^(n-1)/(n-1)!
```

6.5   Indexing Functions

6.5.1   The Purpose of Indexing Functions

The semantics of indexing into a table are described by its indexing function. Using an indexing function, it is possible to do such things as efficiently store a symmetric matrix or count how often each element of a table is referenced. Because each table defines its own indexing method, generic programs can be written that do not need to know about special data representations. For example, the same function could be used to perform an operation on sparse matrices as for dense matrices.

The normal method of indexing, described in Section 6.3, is used when the indexing function of a table is NULL. The semantics correspond roughly to those of common programming languages, with the added notion of "failing" if a component has not been assigned.

If the indexing function for a given table is not NULL, then all indexing into that table is done through a procedure. This procedure is invoked whenever a selection operation of the form T[⟨expression sequence⟩] is encountered and the value of T is the table.

Three parameters are passed to the procedure:

1. the name of the object which is being indexed, T,

2. a list containing the index, [⟨expression sequence⟩],

3. a Boolean value which is **true** (**false**) if the expression is being evaluated as on a left-hand (or right-hand) side of an assignment.

The value returned by the procedure is used in the place of the specified selection operation.

The indexing function may be the procedure itself, or a name. Certain names are known to the basic system as built-in indexing functions. If a name is given, say **IndexingFcn**, which is not one of these then a function call is made using `index/` . **IndexingFcn** . First the current session environment is searched for this name. If it is not found, the Maple system library is searched for the file

`` . libname . `/index/` . IndexingFcn . `.m`

and if no such file exists, then `index/` . **IndexingFcn** is applied as an undefined function.

6.5.2   Indexing Functions Known to the Basic System

Three names are known to the basic Maple system as indexing functions. These are **symmetric**, **antisymmetric**, and **sparse**. Two additional indexing procedures are defined in the Maple library (and hence act as names "known" to the Maple system): **diagonal** and **identity**.

The indexing function **symmetric** is used for tables in which the value of a component is independent of the order of the expressions in the index. The most common application is for symmetric matrices. When a component of a table with this indexing function is referenced, the index expression sequence is re-ordered to give a unique key. (The sort is done using the same internal ordering as for sets.)

Examples:

```
A := array(1..10,1..10,symmetric);    ⟶    A := array(symmetric,1..10,1..10,[ ])
A[1,2];                               ⟶    A[1,2]
A[2,1];                               ⟶    A[1,2]
```

```
A[i,j] - A[j,i];                           ⟶    0
A[3,4] := x;                               ⟶    A[3,4] := x
A[4,3] := y;                               ⟶    A[4,3] := y
lprint(eval(A));                           ⟶    array(symmetric,1..10,1..10,[(3,4)=y])
T := table(symmetric);                     ⟶    T := table(symmetric,[ ])
T[function,continuous,odd] := f;           ⟶    T[function,continuous,odd] := f
T[odd,continuous,function];                ⟶    f
```

The **antisymmetric** indexing function yields the result of **symmetric** multiplied by −1, if all components of the index are different and an odd number of transpositions were necessary to re-order the index. If two or more components of the index are the same, **antisymmetric** returns zero.

Examples:

```
B := table(antisymmetric);                 ⟶    B := table(antisymmetric,[ ])
B[i,j];                                    ⟶    B[i,j]
B[j,i];                                    ⟶    - B[i,j]
B[i,j,k] + B[i,k,j];                       ⟶    0
B[i,k,k];                                  ⟶    0
B[i,j] := v;                               ⟶    B[i,j] := v
B[j,i] := u;                               ⟶    B[j,i] := u
eval(B);                                   ⟶    table(antisymmetric, [(i,j) = - u])
```

The indexing function **sparse** is used with tables for which a component is assumed to have value zero if it has not been explicitly assigned. Suppose T is a table with this indexing function. Evaluating T[⟨expression sequence⟩] on a right-hand side yields the component's value if it has been assigned, or zero if it hasn't. When T[⟨expression sequence⟩] is evaluated on a left-hand side, the indexing function always returns the indexed name T[⟨expression sequence⟩] . since returning zero would make assigning components impossible.

Examples:

```
U := array(1..100,sparse,[90=u1]):
V := array(1..100,sparse,[34=v1]):
s := 0:
for i to 100 do
s := s + U[i] + V[i]
od:
s;                                         ⟶    v1 + u1
```

The two indexing functions defined in the Maple library have the following definitions. The indexing function **diagonal** returns zero as the value of all table components for which the indices are not identical. The indexing function **identity** returns the value one if the indices are all identical, and returns zero otherwise.

6.5.3   User-Defined Indexing Functions

A user may create an indexing function by writing a procedure which returns the expression to be used given the object being indexed, the index, and an indication of whether a left- or right-hand side is desired.

Suppose we wish to define a large tridiagonal matrix. To avoid storing the off-band elements, the following procedure may be used as the indexing function:

```
t3 := proc(A, index, is_lhs)
    local i, j;
    i := index[1];  j := index[2];
    if not type(i-j, integer) or abs(i-j) <= 1 then
            i, j;
            'A["]'
    else
            0
    fi
end;
```

The array may be created by the statement

```
Tri := array(eval(t3), 1..10000, 1..10000);
```

or by the statements

```
`index/tridiagonal` := t3;
Tri := array(tridiagonal, 1..10000, 1..10000);
```

In the first case, `Tri` would have the procedure as its first operand. In the second, it would have the name `tridiagonal`. Assume for the following discussion, that `Tri` has been assigned by the second method.

To explain how this procedure works, suppose the statement

```
Tri[3,20] := rhs;
```

is executed. After the right-hand side has been evaluated, `Tri` is evaluated and found to be an array. Next, the index is evaluated and found to be within bounds. The indexing function is then found to be `tridiagonal` so the following procedure call is made:

```
`index/tridiagonal`('Tri', [3,20], true)
```

The third parameter indicates that the evaluation is being done for the left-hand side of an assignment; in this case the procedure `index/tridiagonal` does not use the third parameter, but it is passed anyway. The element referred to is found not to be on the tri-diagonal band so the **else** part is executed and the value zero is returned. Since it is impossible to make an assignment to zero, the assignment statement generates an error message. This is reasonable; it should not be possible to make assignments to the off-band entries of a tridiagonal matrix.

If the statement

```
Tri[99,100];
```

is executed, then the following events occur. As before, `Tri` is evaluated to a table and the index is found to be within bounds. Then the procedure call

```
`index/tridiagonal`('Tri', [99,100], false)
```

is made. Since this component is found to be on the upper diagonal, the pair of statements

```
i, j;  'A["]'
```

is executed. This returns `Tri[99,100]`, unevaluated, as the procedure value. Then, if `Tri[99,100]` has been assigned, its value is retrieved. Otherwise the table reference fails as usual.

It is important to avoid evaluating the expression `A[99,100]` accidentally inside the procedure, since this would cause an infinite recursion. This is the reason for the unevaluation quotes in the expression `'A["]'` returned by `index/tridiagonal`.

As a second example, suppose we want to count the number of assignments made to components of various arrays and other tables. The counts will be kept in a table, `Count_table`, initialized by

```
Count_table := table(sparse);
```

(Note that `Count_table` needs no explicit initialization of values.) If `A` is one of the tables to be monitored and an assignment is made to `A[1,2]`, then `Count_table[A,[1,2]]` will be incremented by one.

The procedure below may be used as the indexing function for the tables to be monitored:

```
`index/count` := proc(T, index, is_lhs)
    if is_lhs then
        Count_table[T, index] := Count_table[T, index] + 1
    fi;
    op(index);
    'T["]'
end;
```

Then, the tables under investigation are created as follows:

```
t1 := table(count);
a := array(1..100, count);a
```

and used normally.

For a third example, we consider the *Riemann tensor* from general relativity. For our purposes it may be considered to be an array with `(0..3,0..3,0..3,0..3)` as its index bound. This object would have 256 components if all were independent. However, the Riemann tensor has (among others) the following symmetry properties:

```
R[i,j,k,l] = - R[j,i,k,l]
R[i,j,k,l] = - R[i,j,l,k]
R[i,j,k,l] =   R[k,l,i,j] .
```

These imply that at most 21 components are algebraically independent. The array could be created with the following procedure as its indexing function:

```
`index/Riemann` := proc(R,index)
    local i,j,k,l;
```

```
    i := index[1];  j := index[2];  k := index[3];  l := index[4];

    if i = j or k = l then 0
    elif not order(i,j) then -R[j,i,k,l]
    elif not order(k,l) then -R[i,j,l,k]
    elif not order([i,j],[k,l]) then R[k,l,i,j]
    else
         i,j,k,l;
    'R["]'
    fi
end;
```

where **order** is a Boolean function defining an ordering on expressions, such as

```
order := proc(a,b) evalb( a = op(1, {a,b}) ) end;
```

which uses the ordering defined by Maple's ordering of elements in a set.

The procedure `index/Riemann` is recursive, since evaluating the expressions -R[j,i,k,l], -R[i,j,l,k], and R[k,l,i,j] causes the indexing function to be called again.

7
Procedures

7.1 Procedure Definitions

One instance of an expression in Maple is a procedure definition, which has the general form

proc (⟨nameseq⟩) local ⟨nameseq⟩; options ⟨nameseq⟩; ⟨statseq⟩ end.

A procedure definition is normally assigned to a ⟨name⟩ and then it may be invoked using the syntax

⟨name⟩ (⟨exprseq⟩) .

When a procedure is invoked the statements in ⟨statseq⟩ are executed sequentially (and some of the names have special semantics as described below). The *value* of a procedure invocation is the value of the last statement in ⟨statseq⟩ that is executed. For example

```
f := proc(x) x^2 end;        ⟶     f := proc(x) x^2 end
f(2);                        ⟶     4 .
```

Evaluation of an expression in Maple is normally performed in a full, recursive evaluation mode. The result of evaluation is normally the fully evaluated expression. One exception to this rule is when the evaluated expression is a procedure definition, just as the case of tables is an exception. (See Section 6.3 for the evaluation rules for tables.) Namely, if evaluation of an expression yields a procedure definition then the result of evaluation is the *last name evaluated*. The most common situation is the case of a name f which has been assigned a procedure definition, in which case the result of evaluating f will be the name f rather than the value of f. If the value of f is required, the construct

```
eval(f)
```

must be used.

Examples:

```
f := proc(x) x^2 end:
f;                                  ⟶     f
```

```
eval(f);                        ⟶    proc(x) x^2 end
apply := proc(g,a) g(a) end:
apply(f, t);                    ⟶    t^2
apply(eval(f), t);              ⟶    t^2
h := f;                         ⟶    h := f
h(3);                           ⟶    9
```

It is possible in Maple to define and invoke a procedure without ever assigning the procedure definition to a ⟨name⟩. For example:

```
proc(x) x^2 end;                ⟶    proc(x) x^2 end
"(2);                           ⟶    4
proc(x) x^2 end (3);            ⟶    9 .
```

The result of evaluating the first statement above is that the procedure definition is stored and it is available immediately afterwards as the value of the ditto operator. In the second statement, the ditto operator returns the procedure definition as its value and the procedure is then invoked with the argument 2. The third statement is similar to the second, but it expresses the procedure definition and its invocation in a single statement.

Another example of using a procedure definition without a name is when a simple procedure is to be *mapped* onto an expression, as in:

```
a := [1, 2, 3, 4, 5];           ⟶    a := [1, 2, 3, 4, 5]
map( proc(x) x^2 end, a );      ⟶    [1, 4, 9, 16, 25] .
```

The keywords **proc** and **end** may be viewed as brackets which specify that the ⟨statseq⟩ is to remain unevaluated when the procedure definition is evaluated as an expression. The simplest instance of a procedure definition involving no formal parameters, no local variables, and no options can be seen in the following definition of a procedure called **max**:

```
max := proc() if a>b then a else b fi end;
```

Executing the statements

```
a := 25/7;  b := 525/149;  max();
```

yields **25/7** as the value of the procedure invocation **max()**. This procedure is making use of the names **a** and **b** as *global names*. In Maple, all names are global names unless otherwise specified. One instance of non-global names is the case of *formal parameters* which are specified within the parentheses immediately following the keyword **proc**. A more useful definition of the above procedure **max** can be obtained by making the names **a** and **b** formal parameters:

```
max := proc(a,b) if a>b then a else b fi end;
```

This procedure may now be invoked in the form **max(expr1, expr2)** where **expr1** and **expr2** are expressions. For example, **max(25/7, 525/149)** evaluates to 25/7. The names **a** and **b** are now local to the procedure, so that if these names have values external to the procedure the external values neither effect, nor are affected by, the invocation of the procedure.

7.2   Parameter Passing

The semantics of parameter passing are as follows. Suppose the procedure invocation is of the form

$$\langle \text{name} \rangle \ (\ \langle \text{expr} \rangle_1 \ , \ \langle \text{expr} \rangle_2 \ , \ \dots, \ \langle \text{expr} \rangle_n \ ) \ .$$

First, $\langle \text{name} \rangle$ is evaluated and let us suppose for now that it evaluates to a procedure definition with formal parameters

$$\langle \text{param} \rangle_1 \ , \ \langle \text{param} \rangle_2 \ , \ \dots, \ \langle \text{param} \rangle_n \ .$$

Next, the *actual parameters* $\langle \text{expr} \rangle_1 \ , \ \dots, \ \langle \text{expr} \rangle_n$ are evaluated in order from left to right. Then every occurrence of $\langle \text{param} \rangle_i$ in the $\langle \text{statseq} \rangle$ which makes up the body of the procedure is substituted by the value of the corresponding actual parameter $\langle \text{expr} \rangle_i$. It is important to note that these parameters will not be evaluated again during execution of the procedure body; the consequences of this fact are explained in Section 7.6 . In terms of traditional parameter passing mechanisms used by various computer languages, Maple's parameter passing could be termed *call by evaluated name*. In other words, all actual parameters are first evaluated, as in *call by value*, but then a strict application of the substitution rule is applied to replace each formal parameter by its corresponding actual parameter, as in *call by name*.

It is possible for the number of actual parameters to be either greater than, or less than, the number of formal parameters specified. If there are too few actual parameters then a semantic error will occur if (and only if) the corresponding formal parameter is referenced during execution of the procedure body. The case where the number of actual parameters is greater than the number of specified formal parameters is, on the other hand, fully legitimate. If an actual parameter is never referenced it is simply ignored (but it will be evaluated).

Maple allows an alternate mechanism for referencing parameters within a procedure body via the special name **args** which has as its value the expression sequence of actual parameters with which the procedure was invoked. Thus the selection operation **args[i]** references the ith actual parameter. For example, the above procedure **max** could be defined without any specified formal parameters as follows:

```
max := proc() if args[1] > args[2] then args[1] else args[2] fi end;
```

This procedure may now be invoked exactly as before with two actual parameters and the semantics are identical to the previous definition.

In addition to the special name **args**, there are two other special names which Maple understands within a procedure body: **nargs** and **procname**. The value of the name **nargs** is the number of actual parameters (the number of arguments) with which the procedure was invoked. The value of the name **procname** is the procedure name in the sense of the *last name evaluated* (see Section 7.1). If there is no name associated with the procedure then **procname** has as its value the name **unknown**. An example of the use of **procname** can be found in Section 7.7 .

As an example of the use of the name **nargs**, let us generalize our procedure **max** so that it will be defined to calculate the maximum of an arbitrary number of actual parameters. Consider the following procedure definition:

```
max := proc()
    result := args[1];
    for i from 2 to nargs do
        if args[i] > result then result := args[i] fi
    od;
    result
end;
```

With this definition of **max** we can find the maximum of any number of arguments. Some examples are:

```
max( 25/7, 525/149 );        ⟶    25/7
max( 25/7, 525/149, 9/2 );   ⟶    9/2
max( 25/7 );                 ⟶    25/7
max();                       ⟶    Error, (in max) improper op or subscript selector
```

where the latter case is an example of a procedure being called with too few actual parameters. If we wish to change our definition of **max** so that the procedure invocation **max()** with an empty parameter list will return the null value then we may check for a positive value of **nargs** in a selection statement as in the following definition of **max**:

```
max := proc()
    if nargs > 0 then
        result := args[1];
        for i from 2 to nargs do
            if args[i] > result then result := args[i] fi
        od;
        result
    fi
end;
```

Alternatively, one could write the **max** procedure recursively as follows:

```
max := proc(a,b)
    if nargs > 0 then
        if nargs = 1 then a
        elif a > b then max(a, args[3..nargs])
        else max(args[2..nargs])
        fi
    fi;
end;
```

7.3   Local Variables

The mechanism for introducing *local variables* into a Maple procedure is to use the 'local part' of a procedure definition. The 'local part' must appear immediately following the parentheses enclosing the formal parameters, and its syntax is

> **local** ⟨nameseq⟩ ;

The semantics are that the names appearing in ⟨nameseq⟩ are to be local to the procedure. In other words, this can be viewed as causing a syntactic renaming of every occurrence of the specified names within the procedure body.

As an example, let us reconsider the latest definition of `max` appearing above. There are two global variables appearing in the procedure definition which we would almost certainly want to make local: `result` and `i`. This is effected by the following version of the procedure definition.

```
max := proc()
    local result, i;
    if nargs > 0 then
        result := args[1];
        for i from 2 to nargs do
            if args[i] > result then result := args[i] fi
        od;
        result
    fi
end .
```

During the execution of a procedure body, the evaluation of local variables is performed in a *one-level evaluation* mode. This is in contrast to variables which are global to the Maple session, for which full recursive evaluation is applied. The restricted evaluation mode for local variables leads to greater efficiency and the user will find that this evaluation mode is normally sufficient. For exceptions where full recursive evaluation is required, see the `eval` function in the Maple Library.

7.4 Options

There is a facility to specify *options* for a procedure by using the 'options part' of a procedure definition. The 'options part' must appear immediately after the 'local part' and its syntax is either of the following two forms:

> **option** ⟨nameseq⟩ ;
> **options** ⟨nameseq⟩ ;

There are currently seven names which are recognized by Maple when specified in the 'options part' of a procedure definition: `remember`, `builtin`, `system`, `trace`, `operator`, `angle`, and `arrow`.

Option `remember` indicates that the result of any invocation of the procedure is to be stored in a remember table. See Section 7.5 below for a detailed discussion of remember tables in Maple.

Option `builtin` is used to identify Maple's built-in functions. For example, the `type` function is a built-in function in Maple and the result of displaying the value of the name `type` is,

```
eval(type);                    ⟶    proc() options builtin; 126 end;
```

The `builtin` option identifies this as a built-in function rather than a Maple procedure in the external library, and the number 126 is a special number uniquely identifying the particular `type` function.

Option `system` serves to identify procedures which are considered to be *system functions*, meaning functions for which entries may be deleted from the remember table at garbage collection time. If this option is not specified for a procedure having option remember, all entries in the remember table will "survive" garbage collections.

Option `trace` is used to specify that run-time tracing of the procedure should be displayed during execution. The user may explicitly place option `trace` on a procedure to be traced. However, there is a user-interface function called `trace` (see Section 11.2.2) which is often more convenient for this purpose.

Option `operator` identifies the procedure to be a *functional operator*. The final two options, `angle` and `arrow`, only have a meaning when used in conjunction with option `operator`. These options are used by Maple to indicate whether the prettyprinter should print a procedure with option `operator` using angle-bracket notation or using arrow notation.

(See Chapter 8 for a discussion of functional operators in Maple.)

7.5   Remember Tables

With every Maple procedure, there may be associated a *remember* table. A remember table is a hash table in which the arguments to a procedure call are used as the table index, and the result of the procedure call is used as the table value. When a Maple procedure is invoked, its remember table is searched on the actual arguments. If the actual arguments are found as an index in the remember table, the corresponding value is returned as the result of the procedure call. Otherwise the body of the procedure is executed to compute the result. The purpose of remember tables, therefore, is to make use of fast table lookup to avoid expensive recomputation. This is effective when function values are both relatively expensive to compute and likely to be recomputed.

The motivation for remember tables comes from the observation that many operations in symbolic computation reproduce subexpressions several times in their results. All further operations on these results must deal with these common subexpressions repeatedly. By storing the result for each subexpression in the remember table, we avoid recomputation. For example, consider the operation,

```
series( sin(a+x), x=0 );
```
$$\sin(a) + \cos(a)\, x - 1/2 \sin(a)\, x^2 - 1/6 \cos(a)\, x^3$$
$$+ 1/24 \sin(a)\, x^4 + 1/120 \cos(a)\, x^5 + O(x^6)\ .$$

Notice that the resulting series contains many occurrences of the expressions `sin(a)` and `cos(a)`.

Subsequent operations on the resulting series expression will repeatedly encounter these subexpressions.

Values can be entered into a remember table in two ways. First, automatically, by specifying the **remember** option in the options clause of the procedure definition. Second, explicitly, by means of a *function assignment*.

By specifying the **remember** option in a procedure definition, the system automatically records the result of the procedure invocation in the remember table just before returning from the procedure. That is, the actual argument(s) together with the result are inserted as the table index and table value respectively into the remember table.

The *function assignment* allows one to enter results selectively into a remember table. It closely resembles assignment of a table component (Section 6.3.3). It has the form

```
f(x) := y
```
where f must evaluate to a procedure. The meaning of this assignment is as follows. First, the name of the function f is evaluated, then the argument(s) x are evaluated. Next, the right-hand side y is evaluated. Finally, the table entry (x,y) is inserted into f's remember table and the value of y is returned as the result of the assignment. Note that the function call f(x) is not executed.

Values may also be removed from remember tables. The purpose of removing values from a remember table is to free the space which the remember table occupies, and to allow the objects in the remember table to be collected by the garbage collector. If we could not remove entries from a remember table, the problem of running out of memory space could be serious. Entries can be removed from remember tables either explicitly or automatically, as follows. (See also the **forget** function in the Maple Library.)

A procedure's remember table can be accessed as the fourth operand of the procedure. One can delete the entire remember table by substituting the NULL expression in place of it. Alternatively, one can point to it by assigning it to a variable. When accessed in this way, the remember table can be manipulated as a user-level Maple table (see Section 9.3.5). For example, if f is assigned to a procedure, the statements

```
T := op(4, eval(f));
T[1] := evaln(T[1]);
```
remove the value for f(1) from f's remember table (see the **evaln** function in the Maple Library).

By specifying the **system** option in the procedure definition, values are removed from the remember table automatically by the system during garbage collection. To be precise, the values removed are those for which either the argument or the result is not a "live" structure.

Remember tables have proven to be a very useful language feature. Although initially introduced to avoid repeated computations, remember tables have also been used as a programming tool. The remember option allows the coding of linear recurrences and other recursive formulae directly, without a loss of efficiency. For example, the following procedure to compute Fibonacci numbers

```
fib := proc(n)
    if n < 2 then n else fib(n-1) + fib(n-2) fi
end
```

takes exponential time to compute, whereas

```
fib := proc(n) option remember;
    if n < 2 then n else fib(n-1) + fib(n-2) fi
end
```

takes only linear time. Of course, the same improvement could have been obtained either by recoding the procedure explicitly to make use of an array, or by computing the values in a loop. The advantage of **option remember** is that it allows the function to be coded in the most natural manner. The resulting code is very easy to read because the algorithmic intent is not obscured by code for saving intermediate results. A slightly faster version of the Fibonacci program can be obtained by explicitly stating the base of the recursion as follows:

```
fib := proc(n) option remember;  fib(n-1) + fib(n-2)  end;
fib(0) := 0;
fib(1) := 1;
```

Remember tables can be used to tabulate special or exceptional cases in a succinct way by using the functional assignment. They provide direct support for the notion of dynamic programming. In coding complicated recursive algorithms, one can use remember tables to exploit the occurrence of repeated computations, sometimes resulting in a dramatic improvement in running time.

A note of caution. Remember tables should not be used in the presence of side effects. Suppose that one writes a Maple procedure to compute floating-point values for the function f to arbitrary precision. Specifying option **remember** would be incorrect in this case because the precision to which the floating-point values are computed is controlled by the global variable **Digits**. For example, suppose that the value of f(1) is .12345678901234567890 at 20 digits precision. Consider the following:

```
Digits := 10:
f(1);                         ⟶      .1234567890
Digits := 20:
f(1);                         ⟶      .1234567890  .
```

In the second call to f, even though the correct value is .12345678901234567890, the value returned was .1234567890 because .1234567890 was found in f's remember table. The moral is to beware of applying option **remember** with procedures which exploit side effects.

7.6 Assigning Values to Parameters

Let us now consider an example of a procedure where we may wish to return a value into one of the actual parameters. Recall that the integer quotient q and the integer remainder r of two integers a

and b must satisfy the Euclidean division property

$$a = bq + r$$

where either $r = 0$ or $|r| < |b|$. This property does not uniquely define the integers q and r, but let us impose uniqueness by choosing $q = \text{trunc}(a/b)$ using the built-in Maple function trunc. The remainder r is then uniquely specified by the above Euclidean division property. (Note: This choice of q and r can be characterized by the condition that r will always have the same sign as a.) The following definition of the procedure rem returns as its value the remainder after division of the first parameter by the second parameter, and it also returns the quotient as the value of the third parameter, if present:

```
rem := proc(a,b,q)
       local quot;
       quot := trunc(a/b);
       if nargs > 2 then q := quot fi;
       a - quot*b
    end;
```

The procedure rem as defined here may be invoked with either two or three parameters. In either case the value of the procedure invocation will be the remainder of the first two parameters. The quotient will be returned as the value of the third parameter if it appears. At this point recall that the semantics of parameter passing specify that the actual parameters are evaluated and then substituted for the formal parameters. Therefore, an error will result if an actual parameter which is to receive a value does not evaluate to a valid name. It follows that when a name is being passed into a procedure for such a purpose it should usually be explicitly quoted (to avoid having it evaluated to some value that it may have had previously). The following statements will illustrate this concept.

```
rem(5, 2);              ⟶    1
rem(5, 2, 'q');         ⟶    1
q;                      ⟶    2
rem(-8, 3, 'q');        ⟶    -2
q;                      ⟶    -2
rem(8, -3);             ⟶    2
rem(8, 3, q);           ⟶    Error, (in rem) Illegal use of a formal parameter .
```

The latter error message arises because the actual parameter q has the value -2 from a previous statement, and therefore the value -2 is substituted for the formal parameter q in the procedure definition, yielding an invalid assignment statement. The solution to this problem is to change the actual parameter from q to $'q'$.

When values are assigned to parameters within a procedure, a restriction which must be understood is that parameters are evaluated only once. Basically this means that formal parameter names cannot be used freely like local variables within a procedure body, in the sense that once an assignment to a parameter has been made that parameter should not be referred to again. The only legitimate purpose for assigning to a parameter is so that on return from the procedure the corresponding actual parameter has been assigned a value. As an illustration of this restriction, consider a procedure `get_factors` which takes an expression `expr` and, viewing it as a product of factors, determines the number `n` of factors and assigns the various factors to the names `t[1]`, `t[2]`, ..., `t[n]`. Here is one attempt at writing a procedure for this purpose:

```
get_factors := proc(expr,t,n) local i;
    if type(expr, `*`) then
        n := nops(expr);
        for i to n do t[i] := op(i,expr) od
    else
        n := 1;
        t[1] := expr
    fi
end;
```

If this procedure is invoked in the form

```
get_factors(x*y, 't', 'number');
```

the result is

```
Error, (in get_factors) unable to execute for statement .
```

The result occurs because the third actual parameter is a name (as it must be because it is to be assigned a value within the procedure) and when execution reaches the point of executing the **for**-statement, the limit `n` in the **for**-statement is the name **number** that was passed in. Thus the formal parameter `n` is evaluated only once upon invocation of the procedure and it will not be re-evaluated. A general solution to this type of problem is to use local variables where necessary, and to view the assignment to a parameter as an operation which takes place just before returning from the procedure. For our example, the following procedure definition uses this point of view.

```
get_factors := proc(expr,t,n) local i, nfactors;
    if type(expr, `*`) then
        nfactors := nops(expr);
        for i to nfactors do t[i] := op(i,expr) od
    else
        nfactors := 1;
        t[1] := expr
    fi;
    n := nfactors
end;
```

Another solution to the problem in this example is to explicitly invoke the **eval** function (see the Maple Library). The following procedure definition uses this approach:

```
get_factors := proc(expr,t,n) local i;
    if type(expr, `*`) then
        n := nops(expr);
        for i to eval(n) do t[i] := op(i,expr) od
    else
        n := 1;
        t[1] := expr
    fi
end;
```

7.7 Error Returns and Explicit Returns

The most common return from a procedure invocation occurs when execution *falls through* the end of the ⟨statseq⟩ which makes up the procedure body, and the value of the procedure invocation is the value of the last statement executed. There are two other types of returns from procedures.

An *error return* occurs when the special function call

> ERROR(⟨exprseq⟩)

is evaluated. This function call normally causes an immediate exit to the top level of the Maple system. The error header

> Error, (in ⟨procname⟩)

is also printed out, followed by the evaluated sequence of expressions which are parameters to the ERROR function call (typically a single string). Here, ⟨procname⟩ refers to the name of the procedure in which the ERROR return occurred, in the sense of the *last name evaluated* (see Section 7.1). If there is no name associated with the procedure then the name unknown is printed in place of ⟨procname⟩. Errors can be *trapped* by use of the traperror function, and the global variable lasterror stores the value of the latest error message. (See traperror in the Maple Library.)

An *explicit return* occurs when the special function call

> RETURN(⟨exprseq⟩)

is evaluated. This function call causes an immediate return from the procedure and the value of the procedure invocation is the value of the ⟨exprseq⟩ given as the argument in the call to RETURN. In the most common usage ⟨exprseq⟩ will be a single ⟨expression⟩ but a more general expression sequence (including the null expression sequence) is valid. It is an error if a call to the function RETURN occurs at a point which is not within a procedure definition.

A particular form of explicit return is often used as a *fail return*, in the sense that the computation cannot be carried out and it is desired to return the unevaluated function invocation as the result. For this purpose, the two special names procname and args are used, as in

> RETURN('procname(args)') .

Recall that the value of procname is the name of the procedure currently being executed and the value of args is the expression sequence of arguments with which the current procedure was invoked (see Section 7.2). The function invocation appearing here must be quoted to prevent evaluation, for otherwise an infinite recursion will occur.

As an example of a procedure which includes an error return and a fail return, let us consider one final procedure definition for the function max. The latest definition that we developed in Section 7.3 for the function max has a property which makes it unacceptable as a general library function for a computer algebra system, for example. Namely, if a user calls this function with an argument which does not evaluate to a numeric constant, such as in max(x, y) where x and y have not been assigned any values, then the result is an error message from the Maple system: `Error, (in max)` `cannot evaluate boolean`. This error results from attempting to execute an if statement of the form

```
if y > x then . . .
```

where x and y are indeterminates, but the error message is not very informative to the user. In order to improve this situation, type-checking should be done within the `max` procedure and appropriate action should be taken if an invalid argument is encountered. The following procedure definition shows how this can be done using the "fail return" form of an explicit `RETURN`, so that the result of the function call `max(x, y)` will be the unevaluated function `max(x, y)`. This procedure also shows the use of the `ERROR` return facility to return an error message if the function is called with no arguments:

```
max := proc()
    local result, i;
    if nargs = 0 then ERROR(`no arguments specified`)
    else
        result := args[1];
        for i to nargs do
            if not type(args[i], 'numeric') then
                RETURN( 'procname(args)' )
            fi;
            if args[i] > result then result := args[i] fi
        od;
        result
    fi
end;
```

7.8   Simplification and Returning Unevaluated

Often one wants to define transformations that happen for mathematical functions automatically. For example, we may want Maple to recognize that $\sin(n\pi) = 0$ for n an integer, in the sense that whenever $\sin(n\pi)$ is created it gets automatically transformed into 0 . Other identities about the sin function that we may want Maple to know about might include that $\sin(-x) = -\sin(x)$ and $\sin(\arcsin(x)) = x$ for all x . If you like, you can think of these transformations as *simplifications* in the sense that the right hand side of these identities is considered simpler than the left. These kinds of transformations or simplifications are done automatically by Maple. They are provided in the Maple library for the *sin* function and the other elementary functions and special functions i.e. exp, ln, cos, erf, Γ etc. by encoding the simplifications as Maple procedures. The sin function in Maple is initially assigned an ordinary Maple procedure that does the three simplifications above and others. [1] The code has the following structure

---

[1] The actual code for the Maple `sin` procedure, or any Maple library function, may be displayed using the `print` command after first setting the *verboseproc* interface variable to 2. See Section 11.6.

```
sin := proc(x)
    if type(x/Pi,integer) then 0
    elif type(x,`*`) and type(op(1,x),numeric) and op(1,x) < 0 then -sin(-x)
    elif type(x,function) and op(0,x) = arcsin then op(1,x)
    ...
    else 'sin'(x)
    fi
end;
```

An important point that was mentioned in the previous section is what to return from such a procedure if no simplification rules apply, e.g. if the input is just a symbol x. In the above code, if none of the simplification rules apply, the code will end up at the else clause and `'sin'(x)` is returned. We say the procedure returns *unevaluated* because the function call $sin(x)$ is returned as the result. Note that the quotes are clearly needed because without them, the **sin** procedure would be called recursively and would go into an infinite loop. The quotes prevent **sin** from being applied.

We wish to develop two more examples for coding symbolic simplifications to illustrate other kinds of simplifications one may wish to do, and this idea of returning unevaluated if no simplification rules apply. In the first example, we want to define an associative, commutative operator F with identity 0. That is, we want Maple to recognize that $F(x, y) = F(y, x)$, $F(x, F(y, z)) = F(F(x, y), z)$ and $F(x, 0) = x$ respectively. We will write a Maple procedure F that transforms its arguments into a particular form so that commutativity and associativity are recognized. To do this we have chosen to represent F as an n-ary function.

```
F := proc() local a;
    a := [args];
    a := map( proc(x)
                if type(x,function) and op(0,x) = F then op(x) else x fi
            end, a );
    a := sort(a);
    a := map( proc(x) if x = 0 then NULL else x fi end, a );
    if nops(a) = 0 then RETURN( 0 ) fi;
    if nops(a) = 1 then RETURN( op(1,a) ) fi;
    'F'(op(a))
end;
```

Hence we transform $F(a, F(b, c))$ and $F(F(a, b), c)$ into $F(a, b, c)$ to implement associativity, sort the arguments to recognize commutativity, and remove $0's$ recognizing 0 as the identity. The last two special case tests transform $F()$ into 0 and $F(a)$ into a.

Our second example extends our procedure **max** from the previous section to do symbolic simplifications. Our **max** procedure in the previous section returns the maximum of a sequence of numbers of type numeric, i.e. integers or fractions or floating point numbers, but otherwise it simply returns unevaluated. However, we would like our **max** procedure to know that max is commutative and associative, i.e. $max(x, y) = max(y, x)$ and $max(x, max(y, z)) = max(max(x, y), z)$ and also that $max(x, x) = x$, $max(x, -\infty) = x$, and $max(x, \infty)$ is ∞.

```
max := proc() local a;
    a := {args};
    if member( infinity, a ) then RETURN( infinity ) fi;
    a := map( proc(x)
                if type(x,function) and op(0,x) = max then op(x) else x fi
            end, a );
    a := map( proc(x) if x <> -infinity then x fi end, a );
    if nops(a) = 0 then RETURN( -infinity ) fi;
    if nops(a) = 1 then RETURN( op(1,a) ) fi;
    'max'(op(a))
end;
```

Note that by putting the arguments in a set, duplicates are removed and the elements are sorted, thus implementing the simplification rules $max(a, a) = a$ and commutativity.

Our new **max** procedure does not handle numbers. Clearly what one would want to do now is to combine the above symbolic simplifications with the handling of numbers. In addition to this the may want to consider how one might encode more general simplifications for example, $max(\cos(1), \sin(1)) = \cos(1)$, $max(2, \sin(x)) = 2$, and $max(x, x + 1) = x + 1$.

7.9   Boolean Procedures

The names **true**, **false**, and **FAIL** are constants in Maple, and they may also be freely manipulated as names (but they may not be assigned values). It follows that Boolean procedures may be written like any other procedures. As an example of a Boolean procedure, consider the following definition of a function called **member** which tests for list membership:

```
member := proc(element, l)  local i;
    false;
    for i to nops(l) while not " do
        element = op(i, l)
    od;
    evalb(")
end;
```

Some examples invoking this procedure follow:

```
member( x*y, [1/2, x*y, x, y] );   ⟶    true
member( x, [1/2, x*y] );           ⟶    false
member( x, [ ] );                  ⟶    false .
```

In the above procedure, note the use of the nullary operator " in the while-part of the loop to refer to the "latest expression". Alternatively, this could be coded with the use of another local variable. Also note the use of the **evalb** function in the final statement. If the **evalb** function was not applied then the value returned from this procedure would be an algebraic equation

```
element = op(i, l)
```

rather than a Boolean value. During execution of the loop, this equation appears (as the value of ") in a Boolean context and is automatically evaluated as a Boolean. But outside of this Boolean context, it is necessary to force evaluation via the **evalb** function.

7.10   Reading and Saving Procedures

It is usually convenient to use a text editor to develop a procedure definition and to write it into a file. The file can then be read into a Maple session. For example, suppose that the **max** procedure is written into a file with filename **max** (more specifically, the filename on the host computer corresponding to Maple's filename **max**). In a Maple session the statement

        read max;

will read in the procedure definition. Once the procedure is debugged, it is desirable to save it in *Maple internal format* so that whenever it is read into a Maple session the reading is very fast (and no time is spent displaying the statements to the user). To accomplish this, one must use a Maple filename ending in the characters '.m'. Within Maple the user format file is read in and then Maple's **save** statement is used to save the file in Maple internal format. For example, suppose that we have saved our procedure definition as discussed above. If we enter the Maple system and execute the statements

        read max;
        save `max.m`;

we will have saved the internal representation of the procedure in the second file. This file may be read into a Maple session at any time in the future by executing the statement

        read `max.m`;

which will update the current Maple environment with the contents of the specified file. (It is often convenient to place the **save**statement at the end of the user format file so that simply reading in the file will cause it to be saved in Maple internal format.)

A special case of reading procedure definitions in Maple internal format can be accomplished using the built-in function **readlib**. Specifically, the function invocation **readlib(pname)** will cause the following Maple **read** statement to be executed:

        read `` . libname . `/` . pname . `.m`

where **libname** is a global name in Maple which is initialized to the pathname of the Maple system library on the host system. The value of **libname** on any host system can be determined by entering Maple and simply displaying its value. The filename being specified here is a Maple filename which, in general, would be translated into an appropriate filename on the host system. The **convert** function may be used in the form

        convert(⟨Maple filename⟩, hostfile)

to cause Maple to display the actual filename which will be used on the host system.

The complete Maple filename being specified in the above **read** statement is a concatenation of the values of **libname**, '/', **pname**, and the suffix '.m', which could alternatively be specified by

```
cat( libname, `/`, pname, `.m` )  .
```
In order to specify this concatenation using only Maple's concatenation operator '.' it is necessary to concatenate these values to the null string `` `` `` because the left operand of Maple's concatenation operator is not fully evaluated but is simply evaluated as a name. (See Section 3.2.2.)

The **readlib** function has a more general functionality than was indicated above. If it is called with two arguments then the second argument is the complete pathname of the file to be read, and the first argument **pname** is the procedure name which is to be defined by this action. Thus the following two function calls are equivalent:

```
readlib('f')
readlib('f', `` . libname . `/f.m`)
```
but if the procedure definition for 'f' is not in the standard Maple system library then the second argument is required to specify the correct file. Even more generally, the **readlib** function can be called with several arguments, in which case all arguments after the first are taken to be complete pathnames of files to be read, and the first argument is a procedure name that is to be defined by this action. The definition of the **readlib** function involves more than just the execution of one or more **read** statements. This function will also check to ensure that after the files have been read, **pname** has been assigned a value and this value is returned as the value of the **readlib** function. In other words, the **readlib** function is to be used when the purpose of the **read** is to define a procedure named **pname** (and some other names may or may not be defined at the same time).

A common application of the **readlib** function is to cause automatic loading of files. For this purpose, the value of **pname** is initially defined to be an unevaluated **readlib** function, as in one of the following assignments:

```
pname := 'readlib( 'pname' )';
pname := 'readlib( 'pname', filename )';
```
where the single quotes around the argument **pname** are required to avoid a recursive evaluation. Then if there is subsequently a procedure invocation **pname(...)**, the evaluation of **pname** will cause the **readlib** function to be executed, thus reading in the file which defines **pname** as a procedure. The procedure invocation will then continue just as if **pname** had been a built-in function in Maple. In other words, one of the side-effects of evaluating **pname** in this context is that it will be re-defined to have a procedure as its value. Indeed, this method is precisely how the names of Maple's system-defined library functions are initially defined so that the appropriate files will be automatically loaded when needed.

It should be noted that the **readlib** function has option **remember**. Thus a procedure once loaded will not be re-loaded by subsequent calls to the **readlib** function. This fact can be exploited in the sense that a call to **readlib** provides the functionality *load if not already loaded*. Thus a useful format is to load all necessary functions via **readlib**s at the beginning of a procedure body. Every invocation of such a procedure will execute the **readlib** function calls, but these will cost essentially nothing due to option **remember**. (See also the description of **readlib** in the Maple Library.)

8
Operators

8.1 Operator Definition

An *operator* is an abstract data type that describes an operation to be performed on its arguments. This abstract data type is closely associated with the operations of *application* and *composition*, but will also allow most other algebraic operations. The description of the operation may be total, partial, or minimal.

There are several points about operators which will be discussed in this chapter:

- what an operator is,

- how operators are described in Maple,

- how to work with operators in Maple, and

- an example of using an operator .

The various kinds of operators in Maple are shown in Table 8.1 .

Composition is a binary operation, denoted by the symbol @. With this operation, operators form a semigroup with the identity operator (x -> x) as identity element.

Application is a binary operation that takes an operator and a list of arguments and applies the rules that are defined by the operator (if any) to the arguments.

an unassigned name	f
an "ampersand name"	&+, &rank
a procedure	proc(x,y) if y=0 then x^2 else x/y fi end
an "arrow expression"	x -> x^2+sin(x)+1
an "angle-bracket expression"	<x^2+sin(x)+1>
any algebraic combination	exp @ D = 1 + (x -> f(x+1))

TABLE 8.1. Kinds of Maple operators

8.2 Syntactic Definition

Unassigned names in Maple, since there are no declarations, can represent any data type. In particular, they can represent operators.

Another class of operators are the *ampersand names*. Ampersand names are similar to other unassigned names in Maple, except that they have a special syntactic definition. While named operators (such as f) are applied in the form f(x), ampersand operators follow the same syntax as arithmetic operators like +, -, and ⋆; thus they may be used as infix operators.

Ampersand names are formed by the ampersand character '&' followed by a sequence of one or more characters. There are two varieties of &-names, depending on whether the sequence of characters is alphanumeric or non-alphanumeric. In the first case, the symbol consists of & followed by a letter or an underscore, followed by a sequence of zero or more alphanumeric characters (letters, digits, and underscores). For this variety of &-name, it is necessary to use *white space* to delineate the end of the operator name whenever the succeeding operand is alphanumeric; hence, the non-alphanumeric operator names are generally more convenient to use. The second variety of &-name consists of & followed by one or more characters that are non-alphanumeric (so letters, digits, and underscores are excluded) and that are non-punctuation, non-spacing characters (see Section 3.2.7 for a more precise definition).

The &-names can be used as unary or binary operators. For example, &+, &-, and &det are used as binary operators while &inv is used as a unary operator (acting on A):

```
a &+ b &- c &det ( &inv A );      ⟶      ((a &+ b) &- c) &det (&inv A) .
```

"Arrow" operators, or functional operators, are one way to represent a mathematical function, or mapping, in Maple. An arrow operator consists of a sequence of arguments followed by an arrow (->) followed by the expression defining the function. For example,

```
(x, y) -> x^2 + y^2              # takes two arguments and computes
                                 # the sum of their squares

x -> sin(x)                      # defines the sin operator
                                 # (semantically identical to `sin')

x -> x^2 + b                     # squares the argument and adds b

z -> (2*z, z^3)                  # vector function whose first component
                                 # is twice the argument, and second
                                 # component is the cube of the argument .
```

Application of unassigned names and arrow operators uses the *function application* notation; thus the operator appears on the left with a sequence of arguments in parentheses on the right. For example,

```
f(x)
D(sin)
(x -> x^2 + 2*sin(x))(Pi)
```

$$
\begin{array}{ll}
\texttt{(x,y) -> x\^{}2+y} & \equiv \quad \texttt{(z,w) -> w+z\^{}2} \\
\texttt{(x,y) -> x\^{}2+y} & \not\equiv \quad \texttt{(y,x) -> x\^{}2+y} \\
\texttt{x -> x} & \equiv \quad \texttt{y -> y}
\end{array}
$$

TABLE 8.2. Equality between operators

$$
\begin{array}{rcl}
\texttt{a @ (b @ c)} & \equiv & \texttt{(a @ b) @ c} \\
\texttt{a @ (x -> x)} & \equiv & \texttt{a} \\
\texttt{(x -> x) @ a} & \equiv & \texttt{a} \\
\texttt{a @@ 0} & \equiv & \texttt{x -> x} \\
\texttt{a @@ 1} & \equiv & \texttt{a} \\
\texttt{(x -> x) @@ m} & \equiv & \texttt{x -> x} \\
\texttt{a@@m @ a@@n} & \equiv & \texttt{a @@ (m+n)} \\
\texttt{(a @@ m) @@ n} & \equiv & \texttt{a @@ (m*n)} \\
\texttt{a @ (a @@ (-1))} & \equiv & \texttt{x -> x}
\end{array}
$$

TABLE 8.3. Examples of composition of operators

Repeated composition of operators is denoted by `@@`. By definition,

$$
\texttt{a @ a @} \cdots \text{(n times)} \cdots \texttt{@ a} \equiv \texttt{a @@ n} \, .
$$

The `@@` operator has the same precedence as the exponentiation operator and it is not associative. The symbols `@` and `@@` were chosen to emphasize the similarities with the ordinary multiplication and exponentiation symbols `*` and `**`, and to have a composition symbol as close as possible to the circle operator often used in mathematics to represent composition.

8.3 Semantic Definition

Operators, in any of their forms, can be manipulated like any algebraic expressions in Maple. They can be assigned, tested, passed as parameters, and used as arguments in algebraic expressions. Additionally we have the following semantic rules.

Testing equality between operators: Parameters and local variables of operators are considered nameless for comparison purposes. Only their number and positions count (see Table 8.2).

The semantics of *composition* and *repeated composition* can be illustrated by examples as shown in Table 8.3 (where a, b, and c represent operators).

Application of an arrow operator or an angle-bracket operator follows the standard rules of procedure evaluation. These arrow operators are equivalent to single-statement procedures. More precisely,

$$
\texttt{< } \langle \text{expr} \rangle \texttt{ | } \langle \text{params} \rangle \texttt{ | } \langle \text{locals} \rangle \texttt{ >}
$$

$$
\begin{array}{rcl}
\texttt{(a @ b)(x)} & \equiv & \texttt{a(b(x))} \\
\texttt{(a+b)(x)} & \equiv & \texttt{a(x) + b(x)} \\
\texttt{(a*b)(x)} & \equiv & \texttt{a(x) * b(x)} \\
\langle\text{constant}\rangle\texttt{(x)} & \equiv & \langle\text{constant}\rangle \\
\texttt{(x -> x\^{}2+y)(5)} & \equiv & \texttt{25+y}
\end{array}
$$

TABLE 8.4. Examples of operator application

$$
\begin{array}{rcl}
\texttt{x -> a(x)} & \equiv & \texttt{a} \\
\texttt{x -> }\langle\text{constant}\rangle & \equiv & \langle\text{constant}\rangle \\
\texttt{D @ D @ } \cdots \text{ (n times) } \cdots \texttt{ @ D} & \equiv & \texttt{D @@ n} \\
\texttt{(D @ C)(f)} & \equiv & \texttt{D(C(f))}
\end{array}
$$

TABLE 8.5. Automatic simplifications

is semantically equivalent to

$$\langle\text{params}\rangle \texttt{ -> local } \langle\text{locals}\rangle\texttt{; } \langle\text{expr}\rangle$$

and both are semantically equivalent to

proc ($\langle$params$\rangle$) **local** $\langle$locals$\rangle$; **option** operator; $\langle$expr$\rangle$ **end** ;

For the remaining cases, application of operators returns an unevaluated result. For example, if `f` is not defined then `f(x)` returns `f(x)`. Some additional semantics of application are illustrated by examples in Table 8.4.

The function **unapply** is the inverse of application. Application takes an operator and arguments and produces an expression, while **unapply** takes the expression and arguments and will produce an operator (whenever this is possible). The **unapply** function implements the lambda-expressions of lambda calculus. For example,

unapply(sin(exp(x))+cos(x), x); $\longrightarrow$ sin @ exp + cos

and in general

$$\texttt{unapply(} \langle\text{expr}\rangle \texttt{ , x) (x)} \equiv \langle\text{expr}\rangle \texttt{ .}$$

Simplifications are handled automatically (from left to right) as illustrated in Table 8.5 . In addition, simplifications of the form `sin @ arcsin` $\rightarrow$ `(x -> x)` are not performed automatically, but are performed by the function `simplify`.

8.3.1   Application Versus Composition

It is important to emphasize that function application and function composition are not equivalent. This issue will not normally arise unless we are dealing with operators on operators or more complicated cases.

The problem can be stated in the following terms. Let A and B be operators. To resolve any ambiguity, we cannot avoid analyzing the types of the operators, so let the types of A and B be

$$A : X \rightarrow W$$
$$B : Y \rightarrow Z$$

The operators A and B can be said to be of type "mapping", mapping a type X into a type W, and similarly for B. Then the following expressions make sense only if certain conditions hold, as stated:

A @@ n	iff	$X \equiv W$
A @ B	iff	$Z \equiv X$
A(B)	iff	$X \equiv (Y \rightarrow Z)$.

For example, if R defines the domain of computation for sin and cos then

$$\text{sin, cos} : R \rightarrow R$$
$$D : (R \rightarrow R) \rightarrow (R \rightarrow R)$$
$$(x \rightarrow x) : X \rightarrow X \qquad \text{(for any type X)}$$

and therefore

sin @ cos	is valid
sin(a)	is valid if a is of type R
D(sin)	is valid
D @ D	is valid
sin(sin)	is not valid
D(D)	is not valid .

8.4 Partial Definition of Operators

In Maple, the mathematical knowledge about a given function is not all resident in a single piece of code. For example, Maple knows about the function 'sin' in several ways:

sin(Pi);	$\longrightarrow$	0
evalf(sin(ln(2)));	$\longrightarrow$	0.6389612763
series(sin(x), x=0);	$\longrightarrow$	x - (x^3)/6 + (x^5)/120 + O(x^6)
expand(sin(x + Pi/2));	$\longrightarrow$	cos(x)
diff(sin(x^2), x);	$\longrightarrow$	2 x cos(x^2)
evalc(sin(ln(-1)));	$\longrightarrow$	sinh(Pi) I
simplify(sin @ arcsin);	$\longrightarrow$	x -> x .

The philosophy in Maple is to store this knowledge in a distributed form; in this case, in the procedures `sin`, `evalf/sin`, `series/sin`, `expand/sin`, `diff/sin`, `evalc/sin`, and `simplify/@` in contrast to "wiring in" all the knowledge in the functions `expand`, `series`, and so forth. It follows that we can describe some properties of a function without defining the function itself.

The function `define` can be used to define properties of an operator name. It has two basic formats (see the Maple Library for more details).

1. Defining operator names via known abstract algebraic objects. For example,

   ```
   define( group( `&+`, 0, `&-`) );
   ```
   Note that in the define call we want the ampersand names `&+` and `&-` to be treated as ordinary names; the back quotes are required to prevent them from being treated as operators.

2. Defining operator names to have a set of known properties. For example,

   ```
   define(a, associative, identity=0, inverse=b) .
   ```
   Then,

`a(x,0);`	$\longrightarrow$	x
`a(x,b(x));`	$\longrightarrow$	0
`b(0);`	$\longrightarrow$	0
`b(b(x));`	$\longrightarrow$	x
`a(x,b(a(y,x)));`	$\longrightarrow$	y .

The `define` function produces the code for operations such as evaluation, simplification, expansion, differentiation, and series expansion, as needed to handle the operations involving the operator. For example, a matrix multiplication operator could be defined as follows:

```
define(`&@`, binary, identity=Id, associative, zero=0)
```
which will define an operator with the following properties:

`A &@ Id &@ B;`	$\longrightarrow$	`A &@ B`
`evalb(A &@ B = B &@ A);`	$\longrightarrow$	`false`
`A &@ 0 + 3 * Id &@ (3 * B);`	$\longrightarrow$	`9 * B` .

8.5 Example: The Differentiation Operator D

D is a well known and rather special operator and therefore it serves as a useful example. First, note that there are two distinct concepts for *differentiation* in symbolic algebra systems.

1. Differentiate an expression with respect to a variable, which we will denote by `diff`. For example,

`diff(x^2, x);`	$\longrightarrow$	`2*x`

2. A differentiation operator which can be used in algebraic expressions or can operate over functional operators, which we will denote by D. For example,

```
D(sin);                    ⟶    cos
D(x -> x);                 ⟶    1
```

When the derivative cannot be computed the expressions remain unevaluated. For example,

```
diff(f(x), x);             ⟶    diff(f(x), x)
D(f);                      ⟶    D(f)
```

The first derivative of an unknown function f at 0 is expressed as:

```
D(f)(0)
```

and the nth derivative of f at 0 (even if n is undefined) is expressed as:

```
(D @@ n)(f)(0) .
```

For example, a MacLaurin series operator (truncated to six terms) could be defined as

```
ML := f -> local i,x; unapply( sum((D@@i)(f)(0)*x^i/i!, i=0..6), x );
ML(sin);                   ⟶        x -> x - x^3/6 + x^5/120 .
```

Operators are very useful for manipulating differential equations. In mathematics, $(D - a)y$ means $y' - ay$. In Maple, the equivalent construction is achieved with (D-A)(y) where we define the constant-multiplier operator A := x -> a*x . The differential equation $(D^2 - a^2)y = 0$ is expressed in Maple as

```
( D@@2 - A@@2 ) (y)  =  0
```

and one can factor this particular operator in the form

```
D@@2 - A@@2   =   (D-A) @ (D+A)   =   (D+A) @ (D-A) .
```

This is not possible, in general, but in this case the operators D and A are linear operators which commute with each other. The original differential equation becomes

```
( (D+A) @ (D-A) ) (y)  =  0
```

and since D(0) = 0, (D+A)(y) = 0 and (D-A)(y) = 0 are both solutions.

Consider now the differential equation which is commonly written in mathematics using the product notation: $(D-a)^2y = 0$. In Maple notation, we must understand that the relevant operation here is composition of operators, not multiplication, so the proper formulation is

```
( (D-A)@@2 )(y)  =  0
```

or

```
( (D-A) @ (D-A) )(y)  =  0 .
```

This equation has a solution when the first application is equal to 0:

```
(D-A)(y)  =  0                 ⟹    y = C1*exp(a*x) .
```

More generally, there is a solution when the first application gives the result

```
(D-A)(y)  =   C1*exp(a*x)
```

which will make the second application equal to 0, yielding

```
y = C1*x*exp(a*x) + C2*exp(a*x)
```

for arbitrary constants C1 and C2.

9
Internal Representation and Manipulation

9.1　Internal Organization

Maple appears to the user as an "interactive calculator". This mode is achieved by immediately executing any statement entered it the user level. The main program reads input, calls the parser, and then calls the statement evaluator for each complete statement encountered. The parser accepts the Maple language that has been kept simple enough to have the LALR(1) property. For each production which is successfully reduced, it creates the appropriate data structure. Maple will read an infinite number of statements; its normal conclusion is achieved by the evaluation (not the parsing) of the **quit** statement. Thus it is possible to write a statement such as

> **if** ⟨condition⟩ **then quit fi** ;

which will terminate execution conditionally.

The basic system, or *kernel* of Maple, is the part of the system written in a systems implementation language and is loaded in its entirety when a Maple session is initiated. Most of the mathematical knowledge of the Maple system is coded in the user-level Maple language and resides as library functions in the external Maple *library*.　The Maple library functions are described in the *Maple V Library Reference Manual* and in on-line help pages. The functions which are part of the Maple kernel, called internal functions, will be discussed here.

The internal functions in Maple can be divided into four distinct groups.

(1) Evaluators. The evaluators are the main functions responsible for evaluation. There are five types of evaluations: statements (handled by **evalstat**); algebraic expressions (**eval**); Boolean expressions (**evalbool**); name forming (**evalname**); and floating-point arithmetic (**evalf**). Although the parser calls only **evalstat**, thereafter there are many interactions between the evaluators. For example, the statement

```
if a > 0 then b.i := 3.14/a fi;
```

is first analyzed by `evalstat`, which calls `evalbool` to resolve the **if**-condition. Once this is completed, say with a `true` result, `evalstat` is invoked again to do the assignment, for which `evalname` has to be invoked with the left-hand side and `eval` with the right-hand expression. Finally, `evalf` will be called to evaluate the result. Normally, the user will not directly invoke any of the evaluators; these are invoked automatically as needed. In some circumstances, when a non-default type of evaluation is needed, the user can directly call `evalf`, `evalbool` (`evalb`, as called by the user), and `evalname` (`evaln`).

(2) Algebraic functions. These are functions which are directly identified with functions available at the user level, and are commonly called "basic functions". Some examples are: taking derivatives (`diff`), picking parts of an expression (`op`), dividing polynomials (`divide`), finding coefficients of polynomials (`coeff`), series computation (`series`), mapping a function (`map`), substitution of expressions (`subs`, `subsop`), expansion of expressions (`expand`), and finding indeterminates (`indets`). Some functions in this group may migrate to the Maple level (Maple library) and vice versa due to tradeoffs between kernel size and efficiency.

(3) Algebraic service functions. These functions are algebraic in nature, but serve as subordinates of the functions in the above group. In most cases, these functions cannot explicitly be called by the user. Examples of functions in this group are: the arithmetic (integer, rational, and float) packages, the basic simplifier, printing, the series package, the set-operations package, and retrieval of library functions.

(4) General service functions. Functions in this group are at the lowest hierarchical level; that is, they may be called by any other function in the system. Their purpose is general, and not necessarily tied to symbolic computation. Some examples are: storage allocation and garbage collection, table manipulation, internal input/output, non-local returns, and various error handlers.

The flow of control within the basic system is not bound to remain at this level. In many cases, where appropriate, a decision is made to call functions written in Maple and residing in the library. For example, most uses of the function `expand` will be handled by the basic system; however, if an expansion of a sum to a large power is required, then the internal `expand` will call the external Maple library function `` `expand/bigpow` `` to resolve it. Functions such as `diff`, `evalf`, `series`, and `type` make extensive use of this feature. Thus, for example, the basic function `diff` does not know how to differentiate any function; all of its knowledge resides in the Maple library at pathnames `` `diff/⟨function name⟩` ``. This is a fundamental feature in the design of Maple since it permits flexibility (changing the library), personal tailoring (defining your own handling functions), readability (the source is in the user-level Maple code), and it allows the basic system to remain small by unloading non-essential functions to the library.

9.2   Internal Representation of Data Types

The parser and some basic internal functions are responsible for building all of the data structures used internally by Maple. All of the internal data structures have the same general format:

The header field encodes the length of the structure in seventeen bits and the type in six bits, while one bit is used to indicate simplification status, and seven bits are used to indicate which generation of storage the object belongs to, the generation information being used by the garbage collector. The data items are normally pointers to similar data structures.

Every data structure is created with its own length, and this length will not change during its entire existence. Furthermore, data structures are never changed during execution since it is not predictable how many other data structures are pointing to a given structure. The normal procedure for modifying structures is to create a copy and modify the copy, hence returning a new data structure. The only safe modifications are those done by the basic simplifier which produces the same value, albeit simpler. It is the task of the garbage collector to identify unused structures.

In the following figures we will describe the individual structures and the constraints on their data items. We will use the symbolic names of the structures since the actual numerical values used internally are of little interest. The symbol ↑⟨xxx⟩ will indicate a pointer to a structure of type 'xxx'. In particular we will use, whenever possible, the same notation as in the formal syntax (Section 3.3).

Logical And

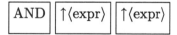

Assignment Statement

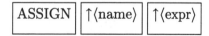

The ⟨name⟩ entry should evaluate to an assignable object, either the NAME, FUNCTION or TABLEREF data structures.

Concatenation of a name

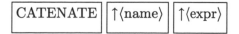

If the ⟨name⟩ entry is one of: NAME, CATENATE, LOCAL, or PARAM, and if the ⟨expr⟩ entry evaluates to an integer or to a name, then the result is a name. If ⟨expr⟩ is a RANGE then the result is to generate an EXPSEQ. (For example, a.(1..2) generates **a1, a2**).

Equation or test for equality

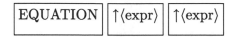

This structure, together with all of the relational operators, has two interpretations depending on the context: as an equation or as a comparison.

Expression sequence

EXPSEQ	↑⟨expr⟩	↑⟨expr⟩	⋯

An EXPSEQ may be of length 1 (no entries); this empty structure is called NULL.

Floating-point number

FLOAT	↑⟨integer⟩	↑⟨integer⟩

The floating-point number is interpreted as the first integer times ten raised to the second integer.

For-while loop statement

FOR	↑⟨name⟩	↑⟨from⟩	↑⟨by⟩	↑⟨to⟩	↑⟨while⟩	↑⟨statseq⟩

or

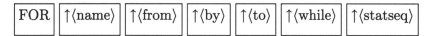

The entries for ⟨from⟩, ⟨by⟩, ⟨to⟩, ⟨in-expr⟩, and ⟨while⟩ are general expressions that are filled with their default values, if necessary, by the parser. The ⟨name⟩ entry follows the same rules as in ASSIGN except that a NULL value indicates its absence. A NULL value in the ⟨to⟩ expression indicates that there is no upper limit on the loop.

Function call

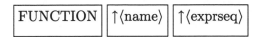

This structure represents a function invocation (as distinct from a procedure definition which uses the PROC data structure). The ⟨name⟩ entry follows the same rules as in ASSIGN, or it may be a PROC definition. The ⟨exprseq⟩ contains the list of parameters.

If statement

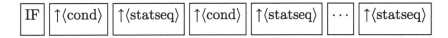

This structure is used to represent **if-then**, **if-then-else**, **if-then-elif**, ... statements. If the length is even, the last entry is an **else**-⟨statseq⟩. The rest of the entries are interpreted in pairs, second and third, fourth and fifth, and so forth. Each pair is a condition (either an **if** or **elif**) and its associated statement sequence.

Not equal or test for inequality

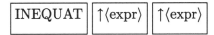

This structure has two interpretations depending on the context: as an inequation or as a comparison.

Negative integer

This data structure is used to represent a negative integer. See the comments below about the representation of integers.

Positive integer

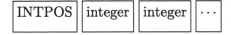

This data structure is used to represent a positive integer. Integers are represented in a base that depends on the host machine (10000 for 32-bit machines). Each entry contains one "digit". A normalized integer contains no additional zeros. The integers are represented in reverse order; the first entry is the lowest order "digit", the last is the highest order "digit". The base used, BASE, is the largest power of ten such that $BASE^2$ can be represented in the host-machine integer arithmetic.

Less than or equal

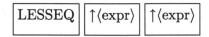

This structure has two interpretations depending on the context: as a relation or as a comparison. The parser also translates a "greater than or equal" into a structure of this type, interchanging the order of its arguments.

Less than

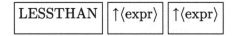

This structure has two interpretations depending on the context: as an inequation or as a comparison. The parser also translates "greater than" into a structure of this type, interchanging the order of its arguments.

List

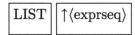

Occurrence of a local variable

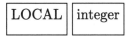

This entry indicates the usage of the ⟨integer⟩th local variable. This structure is only generated by the simplifier when it processes a function definition.

Identifier

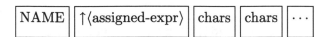

The first entry contains a pointer to the assigned value (if this identifier has been assigned a value) or zero. The chars entries describe the name of the variable. These are null-character terminated, as is the standard in C. The number of characters packed per word is system dependent.

Logical not

Logical or

Occurrence of a parameter variable

PARAM   integer

Similar to LOCAL, but indicating the usage of the ⟨integer⟩th parameter of a function.

Power

POWER   ↑⟨expr⟩   ↑⟨expr⟩

If the second entry is a numeric constant, this structure is changed to a PROD structure by the simplifier.

Procedure definition

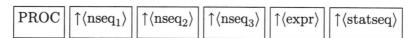

The first ⟨nseq⟩ is an EXPSEQ of the names specified for the formal parameters, the second is an EXPSEQ of the names specified for the local variables, and the third is an EXPSEQ of the options specified. The next location, a pointer to an ⟨expr⟩, is for the procedure's remember table (NULL if there is no remember table). The final location is a pointer to the ⟨statseq⟩ which forms the body of the function.

Product/quotient/power

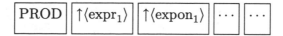

This structure should be interpreted as pairs of factors and their numeric exponents. Rational or integer expressions to an integer power are expanded. If there is a rational constant in the product, this constant will be moved to the first entry by the simplifier.

Range

$$\boxed{\text{RANGE}}\;\boxed{\uparrow\langle\text{expr}_1\rangle}\;\boxed{\uparrow\langle\text{expr}_2\rangle}$$

Rational number

$$\boxed{\text{RATIONAL}}\;\boxed{\uparrow\langle\text{INTPOS or INTNEG}\rangle}\;\boxed{\uparrow\langle\text{INTPOS}\rangle}$$

The second integer is always positive and different from zero or one. The two integers are relatively prime.

Read statement

$$\boxed{\text{READ}}\;\boxed{\uparrow\langle\text{expr}\rangle}$$

The $\langle\text{expr}\rangle$ must evaluate to a name (string) which specifies a file.

Save statement

$$\boxed{\text{SAVE}}\;\boxed{\uparrow\langle\text{exprseq}\rangle}$$

All but the last element in the $\langle\text{exprseq}\rangle$ must be the names of expressions to be saved, and the last element (which may be the only element) must evaluate to a name (string) which specifies a file name.

Series

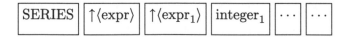

$$\boxed{\text{SERIES}}\;\boxed{\uparrow\langle\text{expr}\rangle}\;\boxed{\uparrow\langle\text{expr}_1\rangle}\;\boxed{\text{integer}_1}\;\boxed{\cdots}\;\boxed{\cdots}$$

The first expression has the general form $\mathtt{x-a}$ where 'x' denotes the variable of the series used to do the series expansion, and 'a' denotes the point of expansion. The remaining entries must be interpreted as pairs of coefficients and exponents. The exponents are integers, not pointers to integers, and appear in increasing order. A coefficient $\mathtt{O(1)}$ (function call to the function "O" with parameter 1) is interpreted specially by Maple as an "order" term.

Set

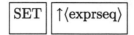

The entries in the ⟨exprseq⟩ are sorted in increasing address order. This is an arbitrary but consistent order, necessary for sets.

Statement sequence

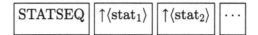

End execution

Sum of several terms

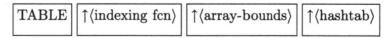

This structure should be interpreted as pairs of expressions and their (integer, rational, or float) factors. The simplifier lifts as many constant factors as possible from each expression and places them in the ⟨factor⟩ entries. An ⟨expr⟩ which is a rational constant is multiplied by its factor and then represented with factor 1.

Table

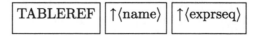

The ⟨indexing fcn⟩ will be either a NAME or a PROC. The next location is a null ⟨exprseq⟩ for general tables. For arrays, it points to an ⟨exprseq⟩ of the array bounds specified as a sequence of integer ranges. The ⟨hashtab⟩ is a hash table storing the table elements.

Table Reference

This data structure represents a table reference, or "indexed name". Its external representation is ⟨name⟩[⟨exprseq⟩]. The ⟨name⟩ entry follows the same rules as in ASSIGN, or it may be a

TABLE structure. (The parser will not generate this structure with a TABLE structure for the ⟨name⟩ entry, but this may happen internally.) The ⟨exprseq⟩ contains the list of indices.

Unevaluated expression

9.3   The Use of Hashing in Maple

Maple's overall performance is in part achieved by the use of table-based algorithms for critical functions. Tables are used within the Maple kernel in both evaluation and simplification, as well as less crucial functions. For simplification, Maple keeps a single copy of each expression or subexpression within an entire session. This is achieved by keeping all objects in a table. In procedures, the **remember** option specifies that the result of each computation of the procedure is to be stored in a "remember table" associated with the procedure. Finally, tables are available at the user level as one of Maple's data types.

All of the table searching is done by hashing. The algorithm is an implementation of direct chaining in which the hash chains are dynamic vectors instead of linked lists. The two data structures used to implement tables are:

Table entry

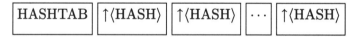

Each entry points to a HASH structure or it is zero if no entry was created. The size of each **HASHTAB** depends on the type of table and on the size of the system. The size of the **HASHTAB** remains fixed during all its existence. To avoid clustering, the number of HASH entries is chosen to be a prime.

Hash chain vector

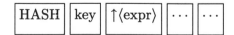

Each table element is stored as a pair of consecutive entries in the hash chain vector. The first entry of this pair is the hash key and the second is a pointer to the stored value. In many cases the key is itself a hashing value (two step hashing). A key cannot have the value zero since this is used as the indicator for the end of a chain. For efficiency, the hash chain vectors are grown a number of entries at a time (proportional to its HASH size) and consequently some of the entries may not be filled.

9.3.1   The Simplification Table

By far, the most important table maintained by the Maple kernel is the simplification table. All simplified expressions and subexpressions are stored in the simplification table. The main purpose of this table is to ensure that simplified expressions have a unique instance in memory. Every expression, which is entered into Maple or generated internally, is checked against the simplification table, and if found, the new expression is discarded and the old one is used. This task is done by the simplifier which recursively simplifies (applies all the basic simplification rules) and checks against the table. Garbage collection deletes the entries in the simplification table which cannot be reached from a global name or from a "live" local variable.

The task of checking for equivalent expressions within thousands of subexpressions would not be feasible if it were not done with the aid of hashing. Every expression is entered in the simplification table using its **signature** as a key. The signature of an expression is a hashing function itself, with one very important attribute: signatures of trivially equivalent expressions are equal. For example, the signatures of the expressions a+b+c and c+a+b are identical; the signatures of a*b and b*a are also identical. If two expressions' signatures disagree then the expressions cannot be equal at the basic level of simplification.

Searching for an expression in the simplification table is done by:

- simplifying recursively all of its components,

- applying the basic simplification rules, and

- computing its signature and searching for this signature in the table.

If the signature is found then a full comparison is performed (taking into account that additions and multiplications are commutative, etc.) to verify that it is the same expression. If the expression is found, the one in the table is used and the searched one is discarded. A full comparison of expressions has to be performed only when there is a "collision" of signatures.

Since simplified expressions are guaranteed to have a unique occurrence, it is possible to test for equality of *simplified* expressions using a single pointer comparison. Unique representation of identical expressions is a crucial ingredient to the efficiency of tables, hence also the **remember** option. Also, since the relative order of live objects is preserved during garbage colle ction, this means that sequences of objects can be ordered by machine address[1]. For example, sets in Maple are represented this way. The set operations union, intersection, etc. can be done in linear time by merging sorted sequences. Sorting by machine address is also available to the user with the **sort** command.

---

[1]Sorting by machine is very fast because a comparison costs one machine instruction.

9.3.2 The Name Table

The simplest use of hashing in the Maple kernel is the *name table*. This is a symbol table for all global names. Each key is computed from the name's character string and the entry is a pointer to the data structure for the name. The name table is used to locate global names formed by the lexical scanner or by name concatenation. It is also used by functions which perform operations on all global names. These operations include:

1. marking for garbage collection,

2. the saving of a Maple session environment in a file, and

3. the Maple functions **anames** and **unames** which return all assigned global names and all unassigned global names, respectively.

9.3.3 Put Tables

It is possible to store Maple objects in a sequential file using a fast-loading internal format. The pointers in a collection of Maple objects form a general directed acyclic graph (DAG). The process of saving values in a file and later reading the values in from the file (usually in a different session) must preserve this graph, and in particular preserve shared subexpressions. A hash table is temporarily created for each **save** or **read** statement that uses internal format. These tables are known in Maple as *put tables*. The put tables are used to keep track of which subexpressions have already been output to (or input from) the file, and, in general, to perform the mapping from a DAG into a linear (labelled) structure.

9.3.4 Remember Tables

Section 7.5 describes in detail the use of remember tables. Here we describe their internal organization, and which internal functions make use of remember tables. A remember table is a hash table in which the argument(s) to a procedure call are stored as the table index, and the result of the procedure call is stored as the table value. Because a simplified expression in Maple has a unique instance in memory, the address of the arguments can be used as the hash function. Hence searching a remember table is very fast.

There are eight kernel functions which use remember tables: **evalf**, **series**, **divide**, **normal**, **expand**, **diff**, **readlib**, and **frontend**.   The internal handling of the latter five is straightforward. There are some exceptions with the first three, namely:

- **evalf** and **series** need to store some additional environment information ('**Digits**' for **evalf** and '**Order**' for **series**). Consequently, the entries for these are extended with the precision information. If a result is requested with the same or less precision than what is stored in the table, it is retrieved anyway and "rounded". If a result is produced with more precision than what is stored, it is replaced in the table.

- `evalf` only remembers function calls (this includes named constants); it does not remember the results of arithmetic operations.

- If the division succeeded and the divisor was a non-trivial polynomial, the `divide` function stores the quotient in its remember table. Otherwise nothing is stored in the remember table.

If option `remember` is specified in conjunction with option `system`, at garbage collection time the remember table entries which would remain, those which are not part of any other "live" expression, are removed. This provides a relatively efficient use of remembering that will not waste storage for expressions that have disappeared from the expression space.

9.3.5   Arrays and Tables in the Maple Language

Arrays and tables are provided as data types in the Maple language. An array is a table for which the component indices must be integers lying within specified bounds. Arrays and tables are implemented using Maple's internal hash tables. Because of this, sparse arrays are equally as efficient as dense arrays.

A table object consists of

1. index bounds (for arrays only),

2. a hash table of components, and

3. an indexing function.

The components of a table T are accessed using a subscript syntax, e.g. `T[a,b*cos(x)]`. Since a simplified expression is guaranteed to have a unique instance in memory, the address of the simplified index is used as the hash key for a component. If no component exists for a given index, then the indexed expression is returned.

The semantics of indexing into a table are described by its *indexing function*. Aside from the default, general indexing, some indexing functions are provided by the Maple kernel. Other indexing functions are loaded from the library or are supplied by the user. Chapter 6 contains a detailed discussion of arrays and tables, including indexing functions.

9.4   Portability of the Maple System

The Maple system is not tied to one operating system. It has been designed to be portable to many operating systems which support a C compiler in a large-address mode (of at least 22 bit addresses). Table 9.1 gives a list of some of the hardware and operating system platforms which currently support a Maple implementation.

To achieve this level of portability and to have a *single* source code (multiple copies are viewed as a disastrous scenario), we use a general purpose macro-processor called Margay. Margay is

Machine	Operating System
386/486-based systems	PC / MS DOS
386-based systems	386/ix, ATT or SCO UNIX
88000-based systems	UNIX
Amiga	Amiga DOS
Apollo 3xxx, 4xxx, DN10000	UNIX
Atari	TOS
AT&T 3B2/500	UNIX V.2
Convex C series	Convex UNIX
Cray 2	Unicos
Data General Aviion	UNIX
DEC Station 2xxx, 3xxx, 5xxx	Ultrix
DEC Microvax	Ultrix or VMS
DEC VaxStation	Ultrix or VMS
DEC Vax	Ultrix or BSD UNIX or VMS
Gould NPI, PN 908x	UNIX
HP 9000/300	HP-UX 7.0
IBM PS/2, Model 70, 80	AIX
IBM RISC System/6000	AIX
IBM RT	AIX
IBM S/370 family	VM/SP CMS
Macintosh Plus, SE, II	Finder
MIPS RSxxxx, RCxxxx, M/2000	UNIX
NCR Tower	UNIX
Pyramid 98xx	Pyramid OSx 4.0/5.0
Sequent Symmetry	Dynix
Silicon Graphics IRIS	IRIX
Sony 68, 3000	UNIX
Sun 3, 4, SPARCstation	SunOS
Sun 386i	SunOS
Unisys Tahoe	UNIX

TABLE 9.1. Some of the platforms supporting Maple

a straightforward macro-processor which resembles closely, although is more powerful than, C's macro-processor. The most important difference is that Margay is written in its own macros and hence it is portable across several systems. The Margay macros reflect machine dependencies, operating system dependencies, size dependencies, and also differences from one C compiler to the next. Another feature of Margay is that it allows functions to have variable-length argument lists.

Once Maple has been ported to a new machine, the user sees a system that is essentially identical to that running on other machines, with a few exceptions such as the character used as the interrupt character, and differences arising from different sizes and speeds.

10
Plotting

10.1 Introduction

An important part of the Maple system is its ability to display functions graphically. This provides the user with a tool for visualizing information about a function. By just glancing at a picture of a function, one can see its zeros, local minima and local maxima, as well other interesting properties. Maple provides the `plot` command and the `plot3d` command for graphing functions of one variable and two variables respectively.

10.1.1 Limitations

In graphing a function by evaluating the function at different points and then interpolating between the evaluation points, you will get a good picture of the behavior of the function most of the time. However, the graph may not always be an accurate reflection of the function's behavior. To get a better picture of a function, you can ask the plotting routines to evaluate the function at more points. This produces a more accurate plot at the cost of more computation time.

But even this has a limit since all plotting programs produce graphs of functions on devices with a fixed resolution, such as a bit-mapped display screen. The amount of information that can be presented is determined by the number of points that the device can display. For example, if the function has a large range within the subdomain between two pixel points, the graph cannot show this information.

Looking at the graph for `sin(x^2)`, one could be led to believe that the function oscillates rapidly and then it stops oscillating at certain regions. You may be lead to believe this for the region near `x = 20` (see Figure 10.1). This would definitely be the wrong impression since the zeroes of this function become more dense as x increases. It just happens that at certain regions, the values of the function at neighboring display points are close together, even though the range of the function in between the two points changes between -1 and 1. This inherent limitation of trying to display continuous information using a finite number of sample points should be kept in mind as one views a graph. Graphs provide only an approximation of the function's behavior, although in most cases it provides a good and very useful approximation.

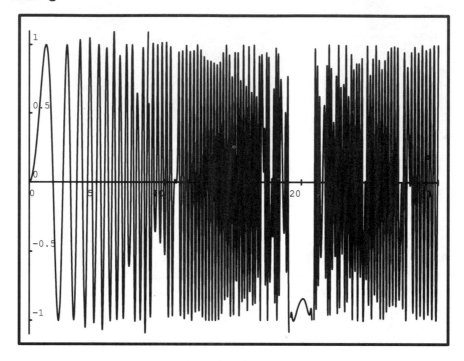

FIGURE 10.1. $\sin(x^2)$ on $x = [0, 30]$

10.1.2 Device Dependencies

Maple has the capability to produce plots on various display and output devices. Many implementations of Maple will choose a display device and a display driver automatically. This is true, for example, for the versions of Maple that run on the X Window System, the Macintosh, the Sunview, and the IBM PC. But for some systems, there is not an obvious choice for a graphics output device and you will be expected to specify one. The Maple command to specify the type of display device is:

```
interface( plotdevice = device );
```

where **device** can be, for example, **char, postscript,** or **tek** for character plots, a Postscript device, or a Tektronix 4010 device, respectively. A complete list of plotting devices that are supported can be seen by asking for the on-line documentation for **plotdevice**:

```
?interface,plotdevice
```

Although many plot drivers are available, some of them cannot meet the demands of producing a 3D plot. For example, on any type of system, you can ask for **plotdevice** to be **char**. This is used to produce graphs on a text terminal by positioning characters over the path of a function. This can be used to generate a very crude plot if you do not have access to a better display device. But it would be a hopeless task to attempt to visualize information from a 3D plot using such a low resolution device. The **plot3d** command will produce an appropriate warning, and not a graph, if the plotting device is not capable of displaying the plot.

10.1.3   Terminal Setup

There are many terminals and terminal emulation programs that operate normally in a text character mode and that can be changed into a graphics mode for graphics output. Sometimes manual intervention is required at the terminal, for example, by pressing special keys or using a setup menu to toggle between text and graphics modes. Often these devices also permit certain character sequences sent to them by a program to switch modes.

Maple provides two interface variables, `preplot` and `postplot`, to be set by a user for the purposes of switching a terminal from text mode to graphics mode and from graphics mode to text mode respectively. Both variables can be assigned a list of integers, corresponding to the integer values of a sequence of ASCII characters. For example,

```
interface( preplot = [ 27, 91, 63, 51, 56, 104 ] );
```

generates a `preplot` string consisting of the characters **escape**, [, ?, 3, 8, and **h**. The `preplot` string is sent to the terminal automatically just before Maple starts drawing the plot. The `postplot` string is sent to the terminal after the plot has been drawn and after the user has pressed the **return** key.

It is sometimes necessary to provide small timing delays as a terminal switches modes before it can be used in the new mode. A negative value $-n$ inserted into the list of integers for `preplot` or `postplot` generates a delay of n seconds.

10.1.4   Hard Copy

When you do not want the graphics output directed to your normal display device, e.g., to your terminal screen, you can have the output saved into a file through the command:

```
interface( plotoutput = file );
```

where `file` is the name of the output file. This file can then be edited, included in other files, or simply printed.

10.2   Plots in 2D

10.2.1   A Simple Example

An example of a graph of a function of one variable is (see Figure 10.2):

```
plot( cos(x), x = 0..2*Pi );
```

This plots the cosine function over the domain $[0, 2\pi]$. Since a range is not given, it will be determined automatically and `plot` will show a graph of the function over the vertical range of $[-1, 1]$. A function to be plotted can be input in one of two forms. It can be an expression of a single indeterminate, or free variable, with the indeterminate specified along with the domain for the graph.

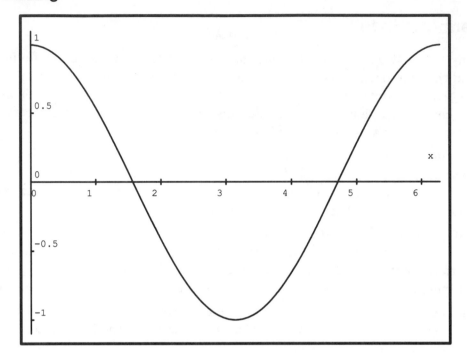

FIGURE 10.2. $\cos x$ on $[0, 2\pi]$

```
plot( exp(cos(x))/x^3, x = 1..10 );
```
Alternatively, it can be input as a functional operator of a single variable. Examples are:
```
plot( f, 0..10 );
plot( proc(x) if x<0 then cos(x) else 1 fi end, -Pi..Pi);
```
where f is a user-defined Maple procedure of a single variable. Using this form, the free variable is not given along with the domain information, and indeed, there is no indeterminate in the operator expression itself. Another example is:
```
plot( cos, 0..2*Pi );
```

10.2.2 Range

To restrict the vertical range of a graph, provide the range information following the domain specification (see Figure 10.3):
```
plot( tan(x), x=0..2*Pi, -5..5 );
```
Parts of the graph outside the vertical range are not shown. Restricting the range instead of allowing Maple to determine the range over the entire domain for the plot is particularly useful when a few values of the function dominate over all the others. In the example above, the tangent function approaches infinity as x approaches $\pi/2$ from the left. Therefore, the range for the graph will be very large if evaluation points are chosen close to $\pi/2$. This has the effect of reducing the vertical scale of the rest of the graph and flattening the display of any details. To counteract this, choose a

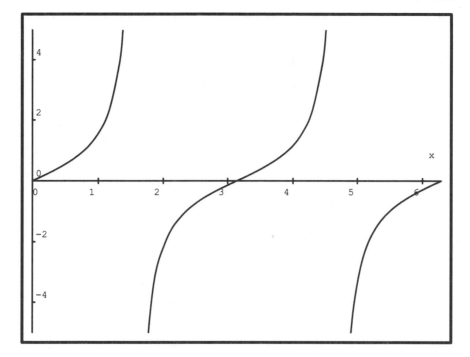

FIGURE 10.3. $\tan x$ on $[0, 2\pi] \times [-5, 5]$

vertical range that prevents singularities from overwhelming the plot.

10.2.3 Style

To generate a graph, Maple samples the function to be evaluated at a number of evaluation points. To draw the curve for the function, it can interpolate values using either linear interpolation or cubic splines. Using splines to interpolate generally produces a smoother looking graph and this is used by default. Specification of `style=LINE` or `style=SPLINE` as one of the options in the `plot` command allows you to explicitly choose one of these methods. Here's an example:

```
plot( exp(x), x=0..2, style=LINE );
```

One other style option allows you to generate a point plot or a scatter plot. The option `style=POINT` stipulates that no interpolating segments are to be drawn and that only each evaluation point is to be shown on the graph.

10.2.4 Parametric Plots

Functions to be plotted may be specified not only explicitly as a function of a single variable but parametrically as a pair of functions of a variable. The first function `x(t)` determines the `x` values of the points to be plotted while the second function `y(t)` specifies the `y` values of the points. An example is (see Figure 10.4):

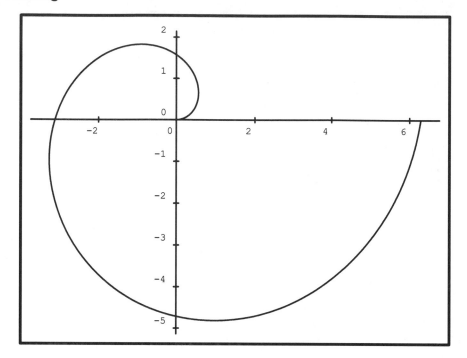

FIGURE 10.4. A spiral. Parametric plot of $(\theta\cos\theta, \theta\sin\theta)$ for θ in $[0, 2\pi]$

```
plot( [ theta*cos(theta), theta*sin(theta), theta=0..2*Pi ] );
```

10.2.5   Polar Plots

If a parametric plot is given and the option **coords=polar** is used, then the parametric functions are interpreted as a pair of functions, radius(t) and theta(t), giving the polar coordinates of the points to be plotted. An example is (see Figure 10.5):

```
plot( [ 1-cos(theta), theta, theta = 0..2*Pi ], coords=polar );
```

10.2.6   Infinity Plots

A novel feature of Maple is its ability to allow you to look at a function over the entire real plane or over a half-plane. The value **-infinity** can be given as the starting endpoint of the domain information and, independently, **infinity** can be given as the final endpoint. An example is (see Figure 10.6):

```
plot( exp(x), x = 0 .. infinity );
```

This gives us an idea of the behavior of the function on the entire right half-plane. It does this by mapping values in $[-\infty, \infty]$ to values in $[-1, 1]$ by using a transformation that approximates **arctan**. Although this fisheye view produces distortions, it gives a convenient overview of the entire

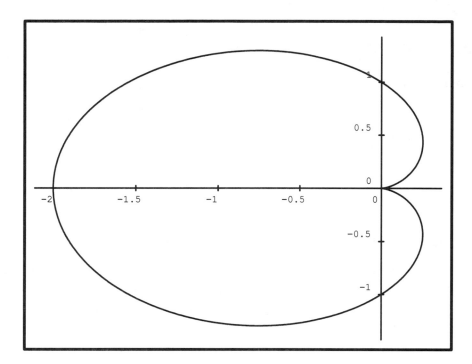

FIGURE 10.5. A Cardioid. Polar plot of $r = 1 - \cos\theta$ for θ in $[0, 2\pi]$

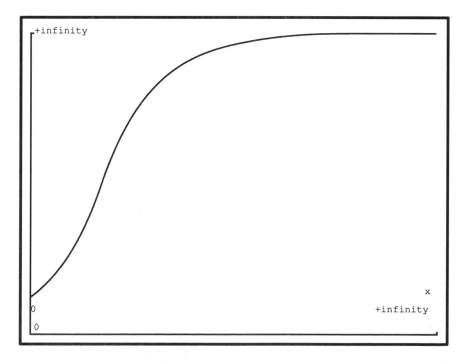

FIGURE 10.6. e^x over $[0, \infty] \times [0, \infty]$

function. If you are plotting a function and you don't know anything about the behavior of the function, an infinity plot helps you pinpoint areas of interest.

10.2.7 Data Plots

Numeric data can also be plotted by Maple. Thus, Maple can be used to produce graphs for values which are collected empirically or which are generated by other programs and read into Maple. A list of points is given by a list of x-y pairs. For example,

```
plot( [5, 5, 6, 6, 7, 8, 9, 20] );
```
plots the x-y pairs $(5,5)$, $(6,6)$, $(7,8)$, and $(9,20)$.

10.2.8 Multiple Plots

It is often useful to compare functions against one another. For this Maple provides the ability to plot several functions at once over a given domain. If the first argument to plot is a set of expressions, instead of a single expression, it will produce a multiplot. Here are two examples:

```
plot( {x-x^3/3, sin(x)}, x = 0..1 );
plot( {x, [x^2, x, x=0..1] }, x = 0..1 );
```
The first example plots two explicit functions. The second one plots one explicit function and one function given parametrically. Where explicit functions and parametric functions are specified together in a multiplot, the same indeterminate variable must be used for each of the functions.

An operator expression containing no indeterminates and an expression over a single indeterminate may not be specified together in a multiplot. For example, this is not allowed:

```
plot( { sin, f, x^2 }, x=0..1 );
```
because x^2 is an expression while sin and f are functional operators. The following can be used:

```
plot( { sin, f, x->x^2 },  0..1 );
```
In the latter example, we have used the functional operator notation for the x^2 function.

10.2.9 Other Options

Options can be specified as a part of the plot command to determine the number of points to be used for evaluation of the function, the number of tick marks to be displayed along the x and y axes, and a title to be added to the graph. Another option tells the plot command the horizontal resolution of the display device. These options are:

- numpoints=n: The minimum number of points at which the function is evaluated. The default value is 49. The initial set of points are spaced uniformly over the graph's domain. An adaptive evaluation method is used to choose more evaluation points in neighborhoods of the function where adjacent evaluation points do not lie close to a straight line.

Color Plates

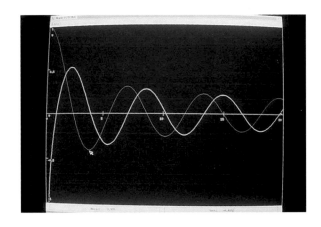

Plate 1. Bessel functions J_0 and Y_0.

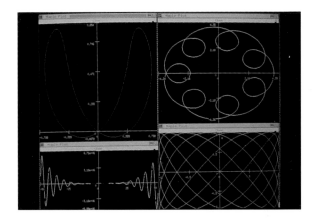

Plate 3. Various 2-d plots.

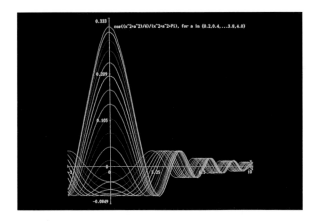

Plate 2. Family of curves (damped cosine curves).

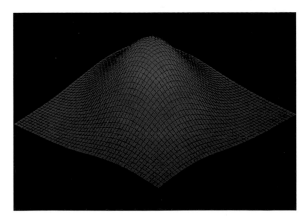

Plate 4. Bivariate Normal Probability Density Function with means (0,0), variances (1,1) and correlation $p = 0.5$.

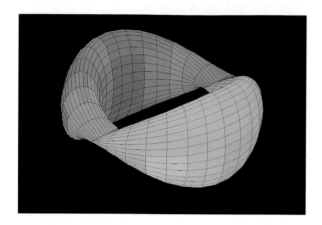

Plate 5. Pinched torus.

Plate 7. Borromean rings.

Plate 6. Spiral tube around a torus.

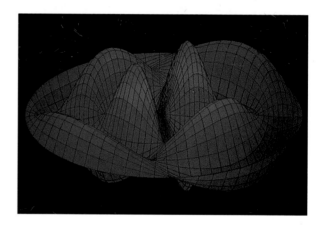

Plate 8. Vibrating drum.

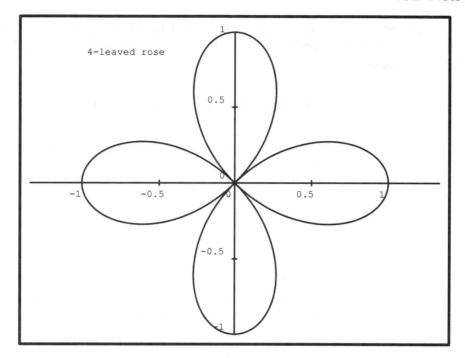

FIGURE 10.7. 4-leaved rose. In polar coordinates $r = \cos 2\theta$ for θ in $[0, 2\pi]$.

- **resolution**=n: The number of points across the display device. This value is used by the adaptive evaluation method to prevent generating more points than can be shown on the display device.

- **xtickmarks**=n: The number of divisions to mark along the horizontal axis.

- **ytickmarks**=n: The number of divisions to mark along the vertical axis.

- **title**=*string*: A title to be added to the graph. The default is no title.

An example showing several different options being used is (see Figure 10.7):

```
plot( [ cos(2*theta), theta, theta=0..2*Pi ], -4/3 .. 4/3, -1..1,
coords=polar, resolution=500, numpoints=200, title=`4-leaved rose` );
```

10.2.10   Common Mistakes

A mistake that is often made is to plot an expression over the variable x when x has already been assigned a value previously. To make this mistake obvious, consider these two lines:

```
x := 1;
plot( x^2, x=0..5 );
```

If the assignment to the variable x had taken place much earlier, and been forgotten, the result for the graph of x^2 could be surprising. Instead of seeing a quadratic curve, you see a straight line.

Since `x^2` evaluates to just 1 as the `plot` function is called, `plot` simply plots the constant function 1. A safe alternative is to guard against evaluation of a variable that you want to be a free variable by calling `plot` in this fashion:

```
plot( 'x^2', 'x'=0..5 );
```

The unevaluation quotes protect `x` from inadvertently being evaluated to a value, leaving it as the name `x` when these arguments are passed to `plot`.

Consider the piecewise function defined by the following Maple procedure

```
f := proc(x)
   if abs(x)<1/100 then x/2-x^3/24
   else (1-cos(x))/x
   fi;
end;
```

It is incorrect to do the following:

```
plot( f(u), u=-3..3);
Error, (in f) cannot evaluate boolean
```

`f` has been evaluated at the value `u`, where `u` has no value it is just the symbol `u`. Again, this is resolved by using quotes to prevent evaluation of the argument 'f(u)' to plot. Hence

```
plot( 'f(u)', 'u'=-3..3 );
```

will work fine. Alternatively the other form of plotting is better in this case:

```
plot( f, -3..3);
```

Another common mistake is mixing operator expressions along with regular expressions with an indeterminate in the function to be plotted. Here is an example of this error:

```
plot( x^2+sin, x=0..Pi);
```

An error message will be produced by Maple noting that there are too many indeterminates in the expression to be plotted.

One final mistake which is often made is attempting to plot an expression that contains symbolic, or unassigned, names. Here's an example of this:

```
plot( [ a*(1-cos(theta)), theta, theta=0..2*Pi ], coords = polar );
```

where `a` has not been assigned a value. The plotting routine will not be able to produce numeric values for the radius as `theta` varies.

10.3   Plots in 3D

10.3.1   Introduction

Maple has a powerful function for displaying 3D graphs of functions of two variables. In addition to being able to plot explicit functions of two variables, it can also plot surfaces defined through parametric functions. Many display options for `plot3d` can be used to generate types of graphs that suit a variety of needs.

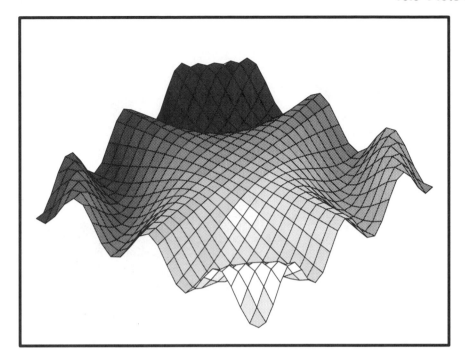

FIGURE 10.8. $\cos xy$ over $[-\pi, \pi] \times [-\pi, \pi]$

On many display devices, once a graph is drawn it is not left just as a bit-mapped image. The graphics object behind the picture is still available and can be actively manipulated. You can change the viewpoint easily. If there is an interesting region of the surface facing away from you, you can grab the surface and rotate it so that the region of interest faces you. You can change the way in which the surface is colored. You can change between a wire-frame drawing style and a style using opaque surface patches.

On many systems, once you have finished examining the graph, you can save it in a file using a number of different graphics formats, including PostScript and troff/pic. These files can be included in documents, or printed.

Some of the functionality just described is system-dependent and these dependencies are described in more detail in the appendices.

10.3.2 Explicit Function Surface Plots

To graph a function of two variables with `plot3d`, you can specify the function in one of two forms. In the first, the expression to be plotted is an expression containing at most two indeterminates, or free variables. The second parameter to the `plot3d` function is an equation giving the name of the variable which varies along the x axis and the domain along that axis. The third parameter gives the same information for the y axis. A simple example is (see Figure 10.8):

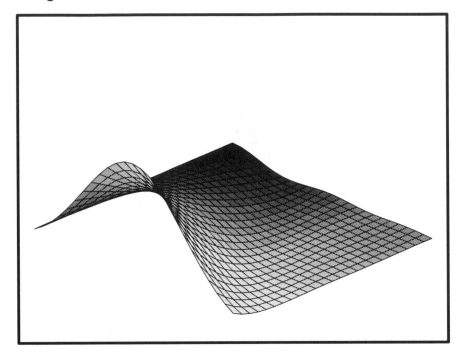

FIGURE 10.9. binomial function over $[0,5] \times [0,5]$

```
plot3d( cos(x*y), x=-Pi..Pi, y=-Pi..Pi, style=PATCH );
```

Using the second form, the expression to be graphed is given as an expression containing Maple procedures and operators of two variables. The expression is not allowed to contain any indeterminate variables. The second argument provides the domain information along the x axis and the third argument provides the same information for the y axis. An example of this is (see Figure 10.9):

```
plot3d( binomial, 0..5, 0..5, style=PATCH );
```

where **binomial** is a Maple function of two arguments.

10.3.3 Parametric Surface Plots

The function to be graphed can also be specified parametrically as a set of three expressions over at most two indeterminates, for example, **s** and **t**. The first expression yields the **x** values for the points to be plotted as **s** and **t** vary. The second expression gives the corresponding **y** values and the third expression gives the **z** values. The names of the two indeterminates must be given along with their domain information. An example of this is (see Figure 10.10):

```
plot3d( [ sin(p)*cos(t), sin(p)*sin(t), cos(p) ],
p = 0..Pi, t = 0..2*Pi, style=PATCH, scaling=CONSTRAINED );
```

Another form for specifying a surface parametrically is to use an operator expression containing Maple procedures and operators taking two variables. The expression must not contain any

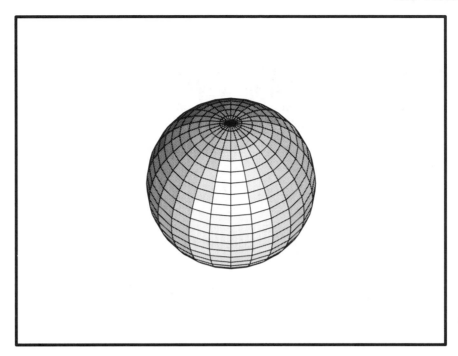

FIGURE 10.10. Unit sphere. Parametrically $(\sin(\phi)\cos(\theta), \sin(\phi)\sin(\theta), \cos(\phi))$ for ϕ over $[0, \pi]$ and θ over $[0, 2\pi]$.

indeterminates and the domain information for the variables of the operators is specified without using names. The following example produces a unit sphere (exactly as in Figure 10.10):

```
f := (p,t) -> sin(p)*cos(t) :
g := (p,t) -> sin(p)*sin(t) :
h := (p,t) -> cos(p) :
plot3d( [ f, g, h ], 0..Pi, 0..2*Pi, style=PATCH, scaling=CONSTRAINED );
```

10.3.4 Style

plot3d can display a graph using different drawing styles according to the option argument style=*DisplayStyle*. The simplest style is POINT, which only draws each of the function evaluation points. A better picture of the graph emerges if line segments are used to connect adjacent points, giving a wireframe drawing, if the style WIREFRAME is chosen. The style HIDDEN goes further and removes all line segments which are hidden behind other parts of the surface. Finally, the style PATCH fills each of the quadrilateral surface patches with an opaque color.

10.3.5 Color

For color display devices, plot3d can specify how the surface is to be colored. An optional parameter for plot3d is shading=*scheme*, where *scheme* is one of XYZ, XY, and Z. Which scheme is used as the default depends upon the display device.

If XYZ coloring is chosen, each of the three axes has its own range of colors and these colors are added together for a point on the surface according to the point's x, y, and z values. Typically, a range of red values are used for one of the axes, a range of green values are used for another axis, and a range of blue values are used for the third axis. The number of colors that are available and exactly what colors are used is device dependent.

If XY coloring is used, the surface is colored using a range of one color along the x axis and a range of another color along the y axis.

Finally, if Z coloring is used, the surface is colored according to the z value of the points on the surface. Typically, blue is used to color low values of z, red is used to color high values, and a mixture of blue and red is used for values in between.

10.3.6   Grid Size

The option `grid=[`m,n`]` specifies the mesh of points over which the function to be graphed is evaluated. The x and y axes are divided uniformly by m and n, respectively. An alternative to the `grid` option is the option `numpoints=`n. If this is used, then the mesh will have $\sqrt{n}$ values along each axis.

If `numpoints` or `grid` is not specified in a call to `plot3d`, then a default mesh size, typically 25 by 25, is used.

10.3.7   Coordinate Systems

The default coordinate system used by `plot3d` is the cartesian (rectangular) coordinate system. The option `coords=`*system*, where the choices for *system* are `cartesian`, `spherical`, and `cylindrical`, allows you to specify different coordinate systems.

10.3.8   Viewpoint

There are several options that affect how the graph is viewed. One is the orientation of the surface to your eyes, i.e., the viewpoint orientation. This can be given in the option `orientation=[` *theta, phi* `]` where *theta* and *phi* are angles, specified in degrees, in the spherical coordinate system. The default orientation if none is given is 45 degrees for each angle. The `view` option allows you to restrict the graph to a certain region. It can be given in two forms. The first `view=`*zmin..zmax* limits the region of the surface to be viewed to the z range between *zmin* and *zmax*. The second form `view=[` *xmin..xmax, ymin..ymax, zmin..zmax* `]` lets you specify a bounding box to be used as the region to which the function should be restricted.

Finally, you can change the amount of perspective transformation that is used in displaying the surface. This is given by the option `projection=`*value* where *value* is a real number in the range

0 to 1. A value of 1 denotes orthogonal projection and a value of 0 denotes a very wide-angle type of perspective transformation.

10.3.9 Scaling of Axes

By default, `plot3d` scales the graph independently along each of the three axes so that it fits the dimensions of the graphics device. Although this maximizes the display of the graph, a 1:1:1 aspect ratio is not guaranteed between the axes. Hence, by stretching to fit, a sphere may be drawn as an ellipsoid. The option `scaling=CONSTRAINED` constrains the scaling so that the 1:1:1 aspect ratio is respected. The option `scaling=UNCONSTRAINED` maximizes the drawing of the graph at the expense of maintaining proper scaling.

10.3.10 Axes and Labels

The option `axes=`*AxesType* specifies whether axes are to be drawn on the graph and if so, how they are to be drawn. The choices for *AxesType* are `NONE`, `NORMAL`, `FRAME`, and `BOXED`. Normal axes pass through the graph and intersect at a point that is usually the origin. Framed axes are drawn along the outer edges of the graph. Boxed axes enclose the entire graph with a bounding box.

The `labels` option allows you to add labels along the x, y, and z axes for the graph. The labels are given as `labels=[` *x_label*, *y_label*, *z_label* `]`, where each of the labels should be an unassigned Maple name, in other words a string. You can also add a title for the graph itself through the option `title=`*a_title*, where *a_title* should be a Maple string.

10.4 Saving Plots

There are different ways of saving the results of a Maple 2D or 3D plot. Both `plot` and `plot3d` generate a Maple data value which is returned as the value of the function call of `plot` and `plot3d`. This plot data structure can be saved in an ordinary Maple variable through an assignment statement.

```
myPlot := plot( sin(tan(x)), x=0..Pi );
```
As is true of any other Maple data value, the plot object can be further assigned to other variables and it can be passed to Maple functions that know how to use them. If you use `lprint` to print the object, you can see exactly what information is maintained as a part of the object. Using the prettyprinter to print the object displays it graphically.

Once computed, Maple plot objects can be saved in a Maple binary file:

```
save myPlot, `results.m`;
```
and then read into another Maple session later:

```
read `results.m`;
```
Finally, versions of Maple for some systems allow you to save the plot in a number of different graphics formats directly from the graphics window.

10.5   Plots Package

Maple has a `plots` package, which provides several utility routines for producing two and three-dimensional plots. Three commonly used procedures in the `plots` package are `matrixplot`, `spacecurve`, and `tubeplot`.

- `matrixplot` defines a three-dimensional graph with the `x` and `y` coordinates representing row and column indices, respectively. The `z` values are given by the input matrix.

- `spacecurve` defines one or more curves in three-dimensional space. Curves are represented by a list of three functions. The three components are considered to be the parametric representations of the `x`, `y`, and `z` coorodinates. `spacecurve` will also accept a set of such lists, in which case several three-dimensional curves are drawn.

- `tubeplot` defines a tube about one or more three-dimensional space curves. You specify the curve parametrically and a radius function. The radius can be constant or a function of the parameters of the curve.

These commands are used to generate some of the plots in the color plate, which we have described in the next section. For more details on the `plots` package, see the *Maple V Library Reference Manual* or the on-line help page, `?plots`.

10.6   Examples

The following examples are included in the color plate. Here we detail the commands used to generate them.

1. Bessel functions J_0 and Y_0.
   ```
   plot({ BesselJ(0,x), BesselY(0,x) }, x=0..20, y=-1..1 );
   ```

2. Family of curves (damped cosine curves).
   ```
   s := {};
   for i from 2 by 2 to 40 do
     a := .1*i;
     s := s union { cos( (x^2+a^2)/4 ) / ( x^2+a^2+Pi ) };
   od:
   plot(s, x=-3..10);
   ```

3. Various 2-D plots.

4. Bivariate Normal Probability Density Function with means $(0,0)$, variances $(1,1)$ and correlation $p = 0.5$.

```
p := .5:
plot3d( 1/ ( 2*Pi*(1-p^2)^(1/2) ) * exp( -1/2*(x^2-2*p*x*y+y^2)/(1-p^2) ),
        x = -2..2, y = -2..2, grid = [50,50]);
```

5. Pinched Torus.

```
plot3d([cos(t)*(1+0.2*sin(u)), sin(t)*(1+0.2*sin(u)), 0.2*sin(t)*cos(u)],
        t = 0..2*Pi, u = -Pi..Pi);
```

6. Spiral Tube Around a Torus.

```
with(plots):
N := 10;
tubeplot({ [10*cos(t), 10*sin(t), 0,
            t=0..2*Pi, radius=2, numpoints=10*N, tubepoints=2*N],
           [cos(t)*(10+4*sin(9*t)), sin(t)*(10+4*sin(9*t)), 4*cos(9*t),
            t=0..2*Pi, radius=1, numpoints=trunc(37.5*N), tubepoints=N] },
        scaling = CONSTRAINED, orientation = [76,40] );
```

7. Borromean Rings.

```
with(plots):
k := 3.0^(1/2);
N := 15;
tubeplot({ [1+k*cos(t), k*sin(t), 0.3*sin(3*t),
            t=0..2*Pi, radius=.3, numpoints=trunc(6.4*N), tubepoints=N],
           [-1/2+k*cos(t), k/2+k*sin(t), 0.3*sin(3*t),
            t=0..2*Pi, radius=.3, numpoints=trunc(6.4*N), tubepoints=N],
           [-1/2+k*cos(t), -k/2+k*sin(t), 0.3*sin(3*t),
            t=0..2*Pi, radius=.3, numpoints=trunc(6.4*N), tubepoints=N] },
        scaling=CONSTRAINED, orientation=[63,37]);
```

8. Vibrating Drum.

```
k := 11.61984117;
plot3d( [ r*cos(t), r*sin(t), BesselJ(2,k*r)*cos(2*t) ],
        r=0..1, t=0..2*Pi, grid=[30,60], orientation=[68,54] );
```

11

Miscellaneous Facilities

11.1   Debugging Facilities: Detecting Syntax Errors

11.1.1   Mint, The Maple Program Diagnostician

The standard mode for detecting syntax errors in Maple procedure definitions is to develop the procedure definition into a file using a text editor external to Maple and then to use the Mint facility to detect any syntax errors. Mint is a program distinct from Maple. It inspects a specified file that is assumed to be a Maple source file and produces a report on syntax errors encountered and on how variables are used.

The precise command syntax for Mint is system-dependent. Generally, Mint processes a specified input file and produces a report. The level of detail presented in the report from Mint can be controlled by an *information level* flag, which takes an integer value between zero and four with the following meanings:

> 0 — display no information
> 1 — display only severe errors
> 2 — display severe and serious errors
> 3 — display warnings as well as severe and serious errors
> 4 — give a full report on variable usage as well as displaying errors and warnings .

The definitions of the various categories of errors and warnings is given below. By default, Mint will use information level 2. A report for each procedure in the input file is displayed separately, followed by a global report for statements not contained within any procedure.

The types of errors and warnings reported are classified as *severe errors*, *serious errors*, and *warnings*. A *severe error* is an indisputable error. A *serious error* is almost certainly an error, but in some circumstances the programmer may choose to ignore it. *Warnings* are possible errors. They point to constructs which may be correct in some contexts, but are probably errors in other contexts. More detailed definitions of these categories follow.

Severe Errors

- *Syntax errors*: A caret symbol (or some other system-dependent symbol) will point to the token which was being read when the error occurred.

- *Duplicated parameter*: A name appears more than once in a parameter list for a procedure.

- *Duplicated local*: A name is declared more than once in the list of local variables for a procedure.

- *Local variable and parameter conflict*: A name is used both as a parameter and as a local variable within a procedure. In further analysis, the name is treated as a parameter.

- *Local variable and system-defined name conflict*: The name of a local variable is also used by Maple as a system-defined name.

- *Parameter and system-defined name conflict*: The name of a parameter is also used by Maple as a system-defined name.

- *Duplicated loop name*: A loop nested within another loop uses as its loop control variable the same name used by the outer loop.

- **break** *or* **next** *statement outside of a loop*: A **break** or a **next** statement occurs outside of any loop. (**break** or **next** may still be used as names within an expression outside a loop.)

- **RETURN** *or* **ERROR** *function call outside of a procedure*: A function call to **RETURN** or **ERROR** occurs outside of a procedure body. (**RETURN** or **ERROR** may still be used as names if they are not invoked as functions.)

- *Unreachable code*: There are statements which follow directly after a "goto" type of statement. These statements are unreachable and will never be executed. A "goto" statement is a **next** statement, a **break** statement, a **quit**, **stop**, or **done** statement, a **RETURN** call, or an **ERROR** call. An **if** statement all branches of which end in a "goto" statement is also considered to be a "goto" statement.

Serious Errors

- *Overly long name*: A name whose length is too long is used. The name is truncated to the maximum allowed.

- *Unused local variable*: A local variable is declared for a procedure but never used within the procedure body.

- *Local variable assigned a value but not used otherwise*: A local variable is assigned a value within a procedure but is not otherwise used.

- *Local variable never assigned a value but used as a value*: A local variable was never assigned a value in a procedure but within the procedure its value is used in an expression. Such an expression would contain a pointer to a non-existent local variable if the expression were returned or assigned to a global variable.

- *System-defined name is overwritten*: A name which is treated as a system-defined name by Maple is assigned a value. Included in the report is an indication of which parts of Maple use the system-defined name.

- *Dubious global name in a procedure*: A global name is used within a procedure. A global name is a name which is not a parameter, a local variable, a system-defined name, nor a concatenated name. Global names used as procedure names in a function call are not considered errors.

- *Library file name overwritten*: The name of a library file, such as `convert/ratpoly`, is assigned a value. It is usual for the name of a library file to be also the name of a library function. Hence, the library function would be no longer accessible.

- *Unused parameter in a procedure*: A name specified in the parameter list of a procedure is never used in the procedure. This is considered a serious error if '**args**' is never used in the procedure either. If '**args**' is used in the procedure, then it is possible that the parameter may be accessed, since **args[i]** accesses the *ith* parameter, and this error is downgraded to a warning.

- *Wrong argument count in a procedure call*: The number of arguments passed in a procedure call does not match the number of formal arguments in the definition of a procedure of the same name recorded in the library database file. The library database file contains information about the minimum number of arguments and the maximum number of arguments expected for a procedure.

Warnings

- *Equation used as a statement*: This may be intentional. On the other hand, it is common for many Fortran and C programmers to mistype '=' for the assignment operator, which is ':=' in Maple.

- *Unused parameter in a procedure*: See similar entry under serious errors.

- *Global name used*: A global name (other than those cases that are *serious errors*) is used within the procedure.

- *Concatenated name used*: A name is formed via the concatenation operator, which implies that it will be a global name.

Other Reports

If the information level is 4, then a usage report is given for each procedure as well as global statements within the file. Each usage report shows how parameters, local variables, global variables, system-defined names, and concatenated names are used. As well as can be done easily, the following information about how a variable is used may be provided:

1. Used as a value.
2. Used as a table or list element.
3. Used as a call-by-value parameter.
4. Used as a call-by-name parameter (a quoted parameter).
5. Called as a function.
6. Assigned a procedure.
7. Assigned a list.
8. Assigned a set.
9. Assigned a range.
10. Assigned a value as a table or list element.
11. Assigned a value to remember as a function value.

In addition, a list of all error messages generated is given.

11.2 Debugging Facilities: Monitoring Run-Time Execution

11.2.1 The printlevel Facility

The printlevel facility allows the user to monitor the execution of statements, to selected levels of nesting within procedures. The default value for printlevel is 1. Any *word-size* integer (on typical host computers, the size limit is nine or ten digits in length) may be assigned to the name printlevel and, in general, higher values of printlevel cause more information to be displayed. Negative values indicate that no information is to be displayed.

More specifically, there are *levels* of statements recognized within a particular procedure (or in the main session) determined by the nesting of conditional and/or repetition statements. If the user assigns

```
printlevel := 0;
```
then the following statements within the main session

```
b := 2;
for i to 5 do a.i := b^i od;
```
will generate the printout b := 2 after execution of the first statement and there will be no printout caused by the **for**-statement, since the value of the **for**-statement is NULL. If the user assigns

```
printlevel := 1;
```
before the above statements are executed, or equivalently, if no assignment to printlevel has been made, then each statement within the **for**-statement will be displayed as it is executed. This would occur in the same manner as if these statements appeared sequentially in the "mainstream", and now will yield the following printouts for the above statements:

```
b := 2
a1 := 2
a2 := 4
a3 := 8
a4 := 16
a5 := 32  .
```

The statement b := 2 is considered to be at level 0 while the other assignment statements in this example are at level 1 because they are nested to one level in a repetition statement. If statements are nested to level i (where **if**-clauses as well as **do**-loops create a nesting of statements within a statement) then the value of `printlevel` must be i if the user wishes to see the results of these statements displayed. Putting it another way, the effective value of `printlevel` is decremented by 1 as execution enters each level of nesting of statements.

More generally, statements are nested to various levels by the nesting of procedure invocations as well. The Maple system decrements the effective value of `printlevel` by 5 upon each entry into a procedure and increments it by 5 upon exit, so that normally (with `printlevel := 1`) there is no information displayed from statements within a procedure. If the user assigns

        printlevel := 5;

in the main session then statements within procedures called directly from the main session (but not nested statements) will be displayed as they are executed, because the effective value of printlevel within the procedure is 0. If the user assigns

        printlevel := 6;

in the main session then, in addition, statements nested to one level in conditional and/or repetition statements in the procedure will be displayed because the effective value of `printlevel` within the procedure is 1. Alternatively, the user may set the value of printlevel explicitly within the procedure for which the information is desired.

It is often useful for debugging purposes to set a high value of `printlevel` in the main session if information is desired from within procedures to various levels of nesting. When the effective value of `printlevel` upon entry to a procedure is 5 or greater, the printout will display the entry point, showing the argument values, and the exit point, showing the returned value, for that procedure. It is not uncommon to use a debug setting such as

        printlevel := 1000;

in which case entry and exit points and statements will be displayed for procedures up to 200 levels deep. For example, consider our `max` procedure from Chapter 7.

```
max := proc() local a;
    a := {args};
    if member( infinity, a ) then RETURN( infinity ) fi;
    a := map( proc(x)
            if type(x,function) and op(0,x) = max then op(x) else x fi
        end, a );
    a := map( proc(x) if x <> -infinity then x fi end, a );
    if nops(a) = 0 then RETURN( -infinity ) fi;
    if nops(a) = 1 then RETURN( op(1,a) ) fi;
    'max'(op(a))
end .
```

The following output occurs when **max** is called as shown. Note the occurrence of the name **unknown** as a procedure name in the following output. This corresponds to invocations of the unnamed procedures which appear as arguments to **map** in the Maple code specified above.

```
> printlevel := 1000:
> max( a, max(b,a) );
--> enter max, args = b, a
                                 a := {a, b}

--> enter unknown, args = a
                                      a

<-- exit unknown = a
--> enter unknown, args = b
                                      b

<-- exit unknown = b
                                 a := {a, b}

--> enter unknown, args = a
                                      a

<-- exit unknown = a
--> enter unknown, args = b
                                      b
<-- exit unknown = b
                                 a := {a, b}

                                   max(a, b)

<-- exit max = max(a,b)
--> enter max, args = a, max(a,b)
                              a := {a, max(a, b)}

--> enter unknown, args = a
                                      a

<-- exit unknown = a
--> enter unknown, args = max(a,b)
                                    a, b
```

```
<-- exit unknown = a, b
                                      a := {a, b}

--> enter unknown, args = a
                                           a

<-- exit unknown = a
--> enter unknown, args = b
                                           b

<-- exit unknown = b
                                      a := {a, b}

                                     max(a, b)

<-- exit max = max(a,b)
                                     max(a, b)
```

When the value of **printlevel** is set to 2 or greater at the top (interactive) level, two other kinds of information are displayed. Firstly, *user-information* printouts will be displayed from procedures which have been programmed to take advantage of this facility. These printouts typically give information about methods (algorithms) being attempted and information about the time taken by and size of various sub-computations. The amount of information is controlled by setting the printlevel variable to 2, 3 or 4 where the higher the number, the more information.

Secondly, if an error return occurs, information about parameters, local variable(s), and the statement being executed in the procedure where the error occurred is displayed. The amount of information is controlled by the setting of the **printlevel** variable to 2, 3 or 4. For example, consider the following Maple procedure **norm** which computes the 2-norm of a univariate polynomial, i.e. the square root of the sum of the squares of the coefficients.

```
norm := proc(a,x) local c,i,s;
    s := 0;
    for i from 0 to degree(a,x) do
        c := coeff(a,x,i);
        s := s + coeff(a,x,i)^2
    od;
    sqrt(s)
end;
```

Then if we call our procedure as follows

```
> norm((x-2)*(x+2),x);
Error, (in norm) unable to compute coeff
```

What went wrong? Running the calculation again with printlevel set to 2 we see

```
> norm((x-2)*(x+2),x);
Error, (in norm) unable to compute coeff
    executing statement: c := coeff(a,x,i)
    locals defined as: c = c, i = 0, s = 0
    norm called with arguments: (x-2)*(x+2), x
```

From this we learn that the input to Maple's **coeff** function has to be *expanded*. That is **coeff** will compute the coefficient of the polynomial $(x - 2)(x + 2)$ only if the input is expanded, i.e. in the form $x^2 - 4$.

11.2.2   Tracing Specific Procedures

A more selective debugging tool is provided by the **trace** function. This function is called with the name(s) of a particular procedure(s) to be traced. Then, during execution, all entry points, with argument values, and exit points, with returned values, and the result of each statement executed inside the procedure will be displayed, for the specified procedure(s). For example, consider our **max** function from the preceding subsection.

```
> trace(max);
> max( a, max(b,a) );
--> enter max, args = b, a
                                a := {a, b}

                                a := {a, b}

                                a := {a, b}

                            max(a, b)

<-- exit max = max(a,b)
--> enter max, args = a, max(a,b)
                          a := {a, max(a, b)}

                            a := {a, b}

                            a := {a, b}

                          max(a, b)

<-- exit max = max(a,b)
                            max(a, b)
```

The general syntax of the trace procedure is

$$\texttt{trace(} \langle f \rangle_1, \langle f \rangle_2, \ \ldots, \ \langle f \rangle_n \texttt{);}$$

where $\langle f \rangle_1$, $\langle f \rangle_2$, $\ldots$, $\langle f \rangle_n$ are names of Maple procedures[1] to be traced. Tracing can be turned off with the **untrace** function which has the syntax

$$\texttt{untrace(} \langle f \rangle_1, \langle f \rangle_2, \ \ldots, \ \langle f \rangle_n \texttt{);}$$

---

[1]Note that unnamed procedures and builtin procedures with special evaluation rules cannot be traced.

11.2.3 Userinfo

Maple provides two facilities for finding out what a calculation is doing, the *printlevel* facility and the *userinfo* facility. The *printlevel* facility, mentioned in a previous section, is intended primarily to be a debugging tool. It essentially provides a trace of the computations done. The *userinfo* facility is intended to inform the user what Maple is doing algorithmically. For example, if the user is doing a large calculation, the user may wish to know how far in the calculation Maple has proceeded. Or, the user may be curious as to how Maple is solving a problem, i.e. what methods are being used.

The userinfo facility works as follows. Programmers insert **userinfo** statements in their programs. A typical example would be

```
userinfo(2, solve, cat(`solving a linear system of `,n,` equations`))
```

If the user performs the assignment

```
infolevel[solve] := 2;
```

then when this userinfo statement is executed, a message like this will be displayed

```
solve/linear:   solving a linear system of 5 equations
```

The **userinfo** statement has the following syntax

$$\texttt{userinfo(} \ \langle\text{level}\rangle\texttt{,} \ \langle\text{topic}\rangle\texttt{,} \ \langle\text{output}\rangle_1\texttt{,} \ \langle\text{output}\rangle_2\texttt{,} \ \dots \ \texttt{)}$$

where $\langle\text{level}\rangle$ is a non-negative integer which determines the level at which information will be printed, $\langle\text{topic}\rangle$ is the name or a set of names of the topic(s) for which this information is to be printed. The information $\langle\text{output}\rangle_1$, $\langle\text{output}\rangle_2$, ... will be printed when the user information statement is executed if the global assignment

```
infolevel[⟨func⟩] := n
```

or

```
infolevel[all] := n
```

has been made and $n \geq \langle\text{level}\rangle$. The user information is displayed as follows. First the name of the invoked procedure is printed followed by a colon and 3 spaces. The information is printed as follows. If an argument is of the form **print(...)** then that argument is prettyprinted. If an argument is of the form **lprint(...)** then that argument is lineprinted on a separate line. Otherwise the arguments are printed together using **lprint** and separated by 3 spaces.

The ability to have different levels allows for different amounts of information to be displayed. The idea is simply, the higher the level, the more information is displayed. The entry **infolevel[all]** is a special entry. If it is assigned n, then every **userinfo** statement will print if its level is less than or equal to n.

Note, that the userinfo facility is a new facility in Maple version V. The functions for which user information is available, and the quality of that information, will be improved and extended in succeeding versions of the system.

11.3   Alias and Macro

Maple provides an *alias* and a *macro* facility whereby expressions can be given labels. A typical example would be

> `alias(J=BesselJ);`

$$I, J$$

> `diff(J(0,x),x);`

$$- J(1, x)$$

This works as follows. In the calculation of the `diff(J(0,x),x)`, the name J is first substituted for the longer unique name **BesselJ** that Maple uses for the Bessel function $J_v(x)$ prior to execution. Then the expression is executed, in this case yielding the derivative, a result in terms of **BesselJ** which is then replaced by J on output. Thus Maple sees and computes with **BesseJ** which it understands, and the user sees and types J .

The *macro* facility is similar to the *alias* facility. It is intended to be used as a programming development tool for naming constants, and as an abbreviation facility. A simple example from one Maple library program is **macro(N=20, M=100)**. The names N and M are used in the programs that follow as if the constants 20 and 100 were used. This makes it very easy to try out different values for these parameters.

11.3.1   Alias

Mathematics is full of special notations and abbreviations. These notations are typically encountered in written material as statements like "let J denote the Bessel function of the first kind" or "let α denote a root of the polynomial $x^3 - 2$". The purpose of the alias facility is to allow the user to state such abbreviations for the longer unique names that Maple uses and, more generally, to give names (or labels) to arbitrary expressions.

The arguments to the **alias** function are equations. When **alias** is called, the equations are evaluated from left to right. The information about the aliases in the system is updated and the result returned is a sequence of all the existing aliases. For example

> `alias( F=f(x), Fx=diff(f(x),x) );`

$$I, J, F, Fx$$

In general, an alias may be defined for any Maple expression except a numerical constant. In our example above, we defined an alias for the expressions `f(x)` and `diff(f(x),x)` such that we can use F and Fx for input and output, for example

> `diff( x*F, x );`

$$F + x Fx$$

In the following, we note some details. On output, Maple has to identify whether the output expression contains any of the alias values, i.e. if it contains `f(x)` or `diff(f(x),x)` in our example. This is done by *substitution* and is thus restricted to the subexpressions that Maple's `subs` function can recognize. For example, if you had defined `alias(T=exp(2*x))`, `exp(x)^2` will not be substituted for `T` in the output even though $exp(2x)$ and $exp(x)^2$ are mathematically equal.

If an alias appears on the left hand side of an assignment statement, the result is that the value of the alias is assigned, not the alias. Thus one should be careful if one mixes assignments with alias. For example

```
> J := 1;
```

$$J := 1$$

```
> BesselJ;
```

$$1$$

The reader will have noticed in our examples that Maple had already defined `I` to be an alias. Maple defines `I` to be an alias for `(-1)^2` . This is Maple's internal representation for the complex unit $\sqrt{-1}$.

When the inputs to `alias` are processed, they are not subject to any existing aliases. This means that you can not define one alias in terms of another. It does mean that the value of an alias can be reset by doing, for example

```
> alias(I=I);
```

$$J, F, Fx$$

or changed by simply redefining it, e.g. `alias(I=BesselI)` . Note, if you did undefine or redefine the alias for `I`, you would have to type `(-1)^(1/2)` for the complex unit. You may want to do this if you want to use a different notation e.g. `alias(j=(-1)^(1/2))` .

In procedures, parameters and local variables are not affected by aliases, although all other expressions are. Thus it is possible to have a local variable or parameter `I` in a procedure even though Maple is using `I` as an alias for the complex unit.

The examples mentioned so far have all shown the use of the alias command at the interactive level. One can also use alias inside a Maple procedure or in a for loop, hence one can define a whole sequence of aliases.

11.3.2 Macro

The `macro` function in Maple works in much the same way as the `alias` function. The main difference is that substitutions happen on input only, and the right hand side can be a numerical constant. Also, unlike `alias`, it does not evaluate it's arguments. Otherwise the way `macro` works is the same as `alias` . It uses *substitution*, and you can undefine and change a macro in the same way. Note the *macro* facility does not support parameterized macros found in some programming languages. It is intended to be a simple abbreviation facility.

Another typical usage of `macro` is to define abbreviations for Maple library functions in a program. For example, if we are writing a Maple procedure that uses the Maple library routine `combinat['fibonacci']` (which computes the fibonacci numbers F_n), we would define F thus `macro(F=combinat['fibonacci'])` .

11.4 Monitoring Space and Time

11.4.1 Run-Time Messages

As execution proceeds in a Maple session, the user will normally see displayed, at regular intervals, a message of the form:

```
bytes used=xxxx, alloc=yyyy, time=zzzz
```

where `xxxx` and `yyyy` are integers, and `zzzz` is a floating-point number.

The value `xxxx` indicates the number of `bytes used`: the number of bytes of memory that have been requested up to that point in the execution of the session. Note that this measure of memory usage is not directly related to the actual memory requirements of the Maple session at any instant, but rather it is a cumulative count of all memory requests made to the internal Maple memory manager during execution of the session. Typically, a significant proportion of the `bytes used` at any time will be re-allocations of actual memory which was previously used and then released to Maple's memory manager. The value `yyyy` indicates the number of bytes of memory *actually allocated* for data space during the Maple session. Note that Maple's internal memory manager requests bytes to be allocated from the host system in large chunks (typically 1024 or 2048 words, where each word is four bytes) and then uses it as needed. The value `zzzz` is the total CPU time for the session, measured in seconds.

The frequency of the `bytes used` messages can be controlled by the user. An implementation of Maple on a particular host system will have a specific *default increment* (such as 1,000,000 bytes used) for these messages. The messages will appear at approximately the frequency specified, but the increment between successive messages will be greater whenever a memory-intensive computation is carried out internally (including the case of loading library files). To change the frequency of the messages, the `words` function is called with argument n, where n is an integer specifying the desired increment. A value of zero is interpreted to mean that the messages should be completely turned off (an 'infinite' increment). See the `words` function in the Maple Library for more information.

The `bytes used` message shown above is also displayed whenever a *garbage collection* occurs, even if the `words` increment has been set to zero. The frequency of garbage collections can also be controlled by the user. Furthermore, it is possible to turn off the `bytes used` messages generated with each garbage collection by setting `gc(0)`. See the `gc` function in the Maple Library.

Finally, the `bytes used` message shown above is displayed when a Maple session is exited via the **quit** (or **done** or **stop**) statement.

11.4.2 The status Variable

The global variable **status** may be queried at any time during a Maple session. The value of this variable is an expression sequence of eight numbers, specifying the following quantities:

status[1]	total number of words requested to the storage allocator
status[2]	total number of words allocated
status[3]	number of CPU seconds used
status[4]	words increment for 'words used' messages
status[5]	words increment for automatic garbage collection
status[6]	number of words returned in last garbage collection
status[7]	number of words available after last garbage collection
status[8]	number of times that garbage collection was executed .

The first three values in the **status** sequence are precisely the values appearing in the **bytes used** message discussed in the preceding subsection, except that for external display there is a conversion from **words** to **bytes** (where one word equals four bytes).

The **status** variable is updated every time that a **bytes used** message is produced and every time that garbage collection occurs. Between these times it remains unchanged. For example,

```
status;      ⟶    1001684, 288768, 126.417, 0, 250000, 132847, 140124, 4
status[5];   ⟶    250000 .
```

11.4.3 System Functions for Space and Time

Three system functions described in the *Maple V Library Reference Manual* and in corresponding on-line help pages are available for measuring and controlling space and time information: **gc**, **words**, and **time**. The **gc** function is used to control the frequency of garbage collections, to explicitly invoke a garbage collection process, and to control the display of the **bytes used** messages caused by garbage collection. The **words** function is used to control the frequency of displaying the **bytes used** messages, and also to query the current value of the **bytes used** measure (see the discussion in Section 11.4.1). The **time** function is used to query the total CPU time used by the session so far, and thus to measure the CPU time used by any selected sub-computation of a session (by querying **time** before and after the selected sub-computation and taking the difference).

There are two miscellaneous functions in the Maple Library which may be used to monitor the space and time usage of a computation: **profile** and **showtime**. The **profile** function, with associated functions **showprofile** and **unprofile**, is used to develop a profile of a function or functions used in a computation, indicating the number of invocations of the function, the nesting level of each invocation, the CPU time used, and the number of **bytes used** by each invocation. The **showtime** function is used to cause the CPU time, and the **bytes used**, to be displayed for each command entered at the interactive level. It also provides an alternate user interface to Maple

name	default	description
constants		sequence of symbolic names defined to be of type `constant`; default sequence: `false`, `gamma`, `infinity`, `true`, `Catalan`, `E`, `Pi`
`Digits`	10	number of digits in floats
`` `mod` ``	`modp`	specifies positive representation for modular arithmetic; to use symmetric, assign `` `mod` `` `:= mods;`
`Order`	6	order of truncation for series
`printlevel`	1	controls run-time printing (see Section 11.2.1)

TABLE 11.1. Global variables with user-controlled values

in which the prompts are labelled I1, I2, I3... , the output values are labelled O1, O2, O3... , and in which previous output values may be accessed by referring to the appropriate `O.i` value (thus providing a complete history mechanism). See also the `history` function in the Maple Library.

11.5 Global Variables

There are five global variables in Maple whose values may be changed by the user to control the computational environment. These are listed in Table 11.1.

Another class of user-controlled variables is related to the user interface and can be set or queried via the `interface` function. These are described in Section 11.6.

In addition to the five global variables listed in Table 11.1 and the 16 user-interface settings described in the following section, there are three additional global variables whose values may be queried by the user. These are listed in Table 11.2. The values of these three variables are not normally changed by the user directly but are set by the system.

11.6 User Interface Variables

The variables described below specify parameters in the user interface. These parameters are set and tested for by the *interface* function. This function is provided as a unique mechanism of communication between Maple and the user interface (called Iris). Specifically, this function is used to set and query all variables which affect the format of the output but do not affect the

name	description
lasterror	the expression sequence which is the latest error message generated in the session (see the **traperror** function)
libname	pathname of the standard Maple library (implementation dependent)
status	expression sequence specifying system status (see Section 11.4.2)

TABLE 11.2. Additional global variables (for user query)

computation.

If the argument is a string, then the current value of that variable is retrieved and returned as the function value. If the argument is an equation then the left hand side must be a string and its value is set to the right hand side of the equation. For example, the call **interface(screenwidth=131);** sets the width of the screen in characters to be 131.

The following is a list of the standard set of interface variables, i.e. the set of variables which should be supported by all types of user interface.

- **echo**: The echo variable controls echoing of input. It must be one of the integers 0, 1 (the default), 2, 3, or 4 where

  0 means do not echo under anything

  1 means echo whenever the input or the output are not from/to the terminal, but do not echo as a result of a read statement.

  2 means echo whenever the input or the output are not from/to the terminal.

  3 means echo only as a result of a read statement

  4 means echo everything

  Note, echoing is superceded by the **quiet** option below, i.e. if **quiet=true**, no echo will occur.

- **quiet**: This is a boolean valued parameter (default **false**) which will suppress all auxiliary printing messages i.e. the Maple logo, garbage collection messages, words used messages, echoing of input, and the prompt.

- **prompt**: This string is printed when input from the user is expected. This happens during normal interactive execution and also by execution of the **readstat** command.

- **screenwidth**: This integer (default 79) specifies the width in characters available on the screen. It must be greater than 10.

- `screenheight`: This integer (default 25) specifies the height in characters available on the screen.

- `prettyprint`: This boolean variable (default **true**) specifies whether the output will be in two dimensions. If **false** it is printed linearly.

- `plotdevice`: This string specifies the name of the plotting device. See help(plot,device) for the known devices.

- `plotoutput`: This string specifies the name of a file where the plot output will be written.

- `preplot`: This list of integers encodes the sequence to be sent before a plot to initialize the plotting device. Negative values specify a delay.

- `postplot`: This list of integers encodes the sequence to be sent after a plot is made to restore the output device. Negative values specify a delay.

- `labelling`: This boolean variable (default **true**) sets and disables % labelling of common subexpressions in large expressions on output (default true). The purpose of the % labelling is to display a large expression more compactly, e.g.

```
> solve(x^3+b*x+c, x);
```

$$\%3 + \%2, \ -\ 1/2\ \%3 - 1/2\ \%2 + 1/2\ 3^{1/2}\ (\%3 - \%2)\ I,$$

$$-\ 1/2\ \%3 - 1/2\ \%2 - 1/2\ 3^{1/2}\ (\%3 - \%2)\ I$$

$$\%1 := (4\ b^3 + 27\ c^2)^{1/2}\ 3^{1/2}$$

$$\%2 := (-\ 1/2\ c - 1/18\ \%1)^{1/3}$$

$$\%3 := (-\ 1/2\ c + 1/18\ \%1)^{1/3}$$

- `labelwidth`: This integer variable (default 20) specifies the minimum size for an object to be considered for a % label.

- `verboseproc`: This variable (default 1) may be assigned the integers 0, 1, or 2. It controls how the contents of a Maple procedure should be printed. If the variable is set to 0, the body of a procedure is not printed. If 1, only user defined procedures are displayed. If 2, the body of all procedures, user and library are displayed. For example,

```
> print(erf);

proc(x) ... end

> interface(verboseproc=2);
```

```
> print(erf);

proc(x)
options `Copyright 1990 by the University of Waterloo`;
    if type(x,float) then evalf('erf(x)')
    elif type(x,numeric) then if x < 0 then -erf(-x) else 'erf(x)' fi
    elif type(x,`*`) and type(op(1,x),numeric) and (op(1,x) < 0) then -erf(-x)
    elif type(x,`+`) and (traperror(sign(x)) = -1) then -erf(-x)
    else erf(x) := 'erf(x)'
    fi
end
```

- **indentamount**: This integer (default 4) specifies how many spaces statements are indented when Maple procedures are printed.

- **endcolon**: This boolean variable indicates whether the last input statement was terminated with a colon (:), instead of a semicolon (;), meaning that the user does not want to see its results. This is a read-only variable.

- **version**: This is a string which provides the name of the interface. It is readable but cannot be set.

11.7   Maple Command Line Options

The command for invoking the Maple system (**maple** on most installations) can take up to three options (fewer on some installations). The details of these options are system-dependent. See the appendices or the system-specific installation notes.

11.7.1   The Quiet Option

This option can be used to start up a Maple session in which no logo will be displayed, no informational messages will be printed, and no sign-off line will be printed. This is particularly useful when the output of a Maple session is to become the input to another process.

11.7.2   The Library Specification Option

This option is used to specify the location of the Maple system library. It is typically used by the system installer when installing the Maple system. On most implementations, users can determine the file system pathname of the Maple library by entering a Maple session and displaying the value of the global variable **libname**.

11.7.3 The Suppress Initialization Option

There is a command line option which may be specified to suppress the loading of initialization files. The concept of initialization files is as follows.

When a Maple session is begun, the Maple system first searches for a *system initialization file* before it starts receiving any input. The location of this file on a particular host system can be determined by the Maple command:

```
convert(``.libname.`/src/init`, hostfile) .
```

There may be no such file specified for a particular implementation, but if the file exists then it is **read** into the session. This file may contain any valid Maple statements, and it is typically used to set parameters for **gc** and **words**.

Secondly, the Maple system searches for a *user initialization file*. The location of this file varies with different host operating systems — see the appendices for system-specific details. This file may or may not exist for a particular user. Any Maple statements may be placed in this file and these statements are executed before proceeding with the Maple session. It is recommended that the statements in this file be terminated with the silent statement terminator ':' if it is desired to have the initialization proceed silently.

11.8 Other Facilities

11.8.1 Executing Other Programs: The system Command

The **system** command is used to run a command in the host operating system. This command takes one argument, a **string**, and executes the command in that string in the host operating system (for example Unix, CMS, VMS etc.) and returns to Maple the *exit status* of the command.

The **system** command allows a Maple program to execute another program on the host system. A typical application of the system command would be to execute a C or Fortran program, possibly on some data which was generated inside Maple, and have the C or fortran program put the result in another file which could then be read back into Maple. The **system** command therefore allows Maple programs to execute other system programs and communicate with them via files. Data can be output from Maple to a file using the **writeto** or **appendto** functions. Data can be read back into Maple from a file using the **read** statement. For example, under the Unix operating system, the call

```
system(`approx < mapout > mapin`);
```

runs the system program **approx** with input from the file **mapout** and output going to the file **mapin**. Note that if the output is going to be read into Maple, it must be readable by Maple, i.e. the syntax must correspond to Maple's syntax.

The *exit status* of the **system** command depends on the operating system. It is usually an integer which indicates whether the command executed successfully, or whether it terminated because of an error. For example, under the Unix operating system, if a command executes and returns normally, the *exit status* returned is the integer 0.

Finally note that on some operating systems, some operating system commands may not be accessible via this mechanism.

11.8.2   Escape to Host

The exclamation mark character ! when it appears as the first character on a line is treated as an "escape to host" operator. All characters following it are given to the host system as a command to be executed. The line need not terminate with a Maple statement separator (':' or ';') before the *return* character terminates the line. The ! character is shorthand for the **system** function described in the previous subsection. It is intended primarily for interactive usage.

11.8.3   Help

The character ? when it appears as the first character in a line is treated as an "escape to **help**" operator. Specifically, the word(s) following the ? character are passed to Maple's on-line **help** facility. This provides a user interface to the **help** facility which is more user-friendly than the standard *function call* syntax of the **help** procedure. The line need not terminate with a Maple statement separator (':' or ';') before the *return* character terminates the line. Furthermore, special names such as keywords or global variables which would need some form of quotes when passed to the **help** procedure explicitly, do not require quotes in the context of the ? syntax.

To get started using Maple's on-line **help** facility, simply type ? followed by *return*.

11.8.4   Interactive Input

The **readstat** function allows a program to prompt the user for interactive input. The argument to the readstat command is a string which is used as the prompt. The user then enters a valid Maple expression, terminating it with a Maple statement separator (':' or ';'), after which the program resumes execution. For example:

```
> f := proc() local n;
>      n := readstat(`Input the degree> `);
>      while not type(n,posint) do
>           n := readstat(`Input the degree (it must be a positive integer)> `);
>      od;
>      factor(x^n-y^n)
> end:
> f();

Input the degree> N;
Input the degree (it must be a positive integer)> 6;
```

$$(x - y) (x + y) (x^2 + x y + y^2) (x^2 - x y + y^2)$$

11.8.5   File Input and Output

Text files containing Maple statements can be read into Maple using the **read** statement – see Section 3.1.3. The file will be read and the Maple statements in it will be executed. The data in the file must be syntactically valid Maple input. It is not possible in the present version of Maple to read arbitrarily formatted data.

The **appendto** and **writeto** functions are used to redirect Maple output to a given file. Output normally would go to the user's terminal. Output from Maple is typically generated at the interactive level as the result of normal execution, and also from executing the Maple functions **lprint** and **print** .

The **appendto** and **writeto** functions take one argument, a string, which is the name of the file. After executing them, all output from Maple will be directed to this file. The output can be redirected back to the terminal by executing **appendto** or **writeto** on the string **terminal**. The **appendto** function differs from **writeto** in that output will not overwrite what is already in the file, rather, it *appends* the output to the file. If the **writeto** function is used, output overwrites the file.

After output has been directed to a file, the Maple functions **lprint** and **print** can be used to print values into that file. The **cat** function for string and integer concatentation is be useful for creating arbitrary text to be printed.

Note that normal Maple output also includes diagnostic messages such as the *status message*, and the *prompt* and input line depending on the setting of the interface variable **echo**. In most cases you will not want this output in the file. You can turn off the printing of this output by setting the *quiet* option using the **interface** function as follows:

```
interface(quiet=true);
```

See Section 11.6 on the **interface** function for how to reset the *quiet* option to its previous value.

12
Overview of the Maple Library

12.1 Introduction

The *Maple V Library Reference Manual* describes all the functions available in the Maple library. The library is where the code for over 95% of Maple's mathematical knowledge and expertise resides. The library routines are available for your inspection, study, and when appropriate, modification and extension. Each of these routines is written in the Maple programming language. Through the library, the Maple system and the algorithms that it uses can be dynamically extended and customized.

Every Maple library routine is described in detail in the *Maple V Library Reference Manual*. Each description includes a brief description of the function, the parameters it uses, and further details about the function. Also included are extensive examples of the use of each routine. One of the easiest ways to become familiar with a new routine is to examine some typical uses of the routine.

All of the descriptions of the Maple library functions can also be produced on-line from within a Maple session. Once Maple has been invoked, the command ?*topic* will show you the documentation for the topic. For example, ?`linalg` will produce the same documentation for the `linalg` (Linear Algebra) package of routines as is printed in the *Maple V Library Reference Manual*.

The electronic form of these descriptions does have one advantage over the printed form. With many versions of Maple, especially for workstations, the documentation appears as a separate window on the screen. Electronic cutting and pasting operations allow you to copy an on-line example of a function's use and paste it into your main Maple worksheet. You may then execute the example immediately or you may edit the example and then execute it. This process is an easy way to explore the capabilities of new Maple functions that you wish to learn.

12.2   Description of the Maple Library

Categories of Library Functions

Maple's library functions fall into four categories: 1) "builtin" functions internal to the Maple system, 2) demand-loaded library functions, 3) miscellaneous library functions which are not automatically loaded on demand, and 4) packages of functions which are normally loaded via the `with` command.

Functions in the first category are coded in the Maple kernel using the system implementation language to achieve efficiency and in some cases to achieve functionality which would not be available via user-level Maple code. Functions in the second category are specified by Maple code in the Maple system library, and their names are initially assigned as unevaluated `readlib`-defined functions (see Section 7.9). Functions in the third category are specified by Maple code and they cannot be used without being explicitly loaded by the user, using the `readlib` function. Packages of functions are specified by Maple code and may be loaded via the `with` library function, as described below.

The distinction between the first two categories will be essentially irrelevant to Maple users. For a specific function 'f', the user can easily determine which of the two categories it belongs to by entering Maple and using the command `eval(f, 1);`. This command evaluates `f` just one level; this allows us to peek into the contents of the variable `f` without evaluating it fully.

For functions in the first category, the result will be a procedure definition with '**options builtin**' specified and with an integer for the procedure body, as in

```
proc () options builtin; 51 end
```

This indicates that this function is built into the Maple kernel and is coded in the system implementation language and not in the Maple programming language. Builtin functions cannot be inspected by the user since only Maple language routines can be meaningfully displayed. The number of builtin functions in the library is minuscule compared with the rest of the library, which can be inspected.

For functions in the second category, the result of `eval(f, 1);` will be

```
readlib('f')
```

This shows that the function `f` is defined in the library. By noting that `f` is `readlib`-defined, we mean that the initial value of the variable `f` is the unevaluated `readlib` invocation shown. As soon as `f` is evaluated or used normally (by being part of a function call), the library Maple-level code for `f` is automatically loaded and assigned to `f`. Such automatic demand-loading of code for the majority of Maple functions gives the user instant and transparent access to the library. Demand-loading provides the advantage of requiring space only for code that is actually used.

There are miscellaneous library functions which are available in the Maple library but which must be explicitly loaded before they can be used. Since these functions are not `readlib`- defined, before using one of them you must load it explicitly through a command such as

```
readlib(f);
```
This reads the procedure definition for **f** from the appropriate place in the library and then assigns the procedure to **f**.

Inspecting Maple Library Routines

With the exception of the builtin functions, the Maple source code for each routine may be displayed. To look at the code for a function **f**, use the command

```
print(f);
```
The function **f** is evaluated normally before being printed. If it had not been loaded previously and is **readlib**-defined, the code for **f** will be loaded. If the function **f** is one of the miscellaneous library functions in category 3 and therefore is not **readlib**-defined, you will need to explicitly load the code for **f** before printing it.

The code displayed for **f** may be in either an abbreviated form or a full display form. The abbreviated form simply displays the function header for **f**, followed by ... for the deleted body, followed by **end**. The full display form shows the entire definition of the procedure in a nicely indented fashion. A user interface setting allows the user to choose how procedures are to be displayed. The command

```
interface( verboseproc=setting );
```
is used to set **verboseproc** to 0, 1, or 2. By default, the value for **verboseproc** is 1. The value of 1 prints user-defined procedures in full display form and library procedures in abbreviated form. A setting of 2 displays both in full display form. A setting of 0 displays both in abbreviated form.

The code which is displayed is formatted by Maple from the internal code that is loaded for the procedure. Since Maple source level comments are not maintained in the internal code, you will not see these nor will you see the code as it was originally formatted in the Maple source files. All of the original, commented Maple source code for the library can be ordered by contacting:

Waterloo Maple Software
160 Columbia Street West
Waterloo, Ontario
Canada N2L 3L3
Telephone: (519) 747-2373
E-mail: wmsi@daisy.uwaterloo.ca

Argument Evaluation Rules

For all functions, all arguments are evaluated (from left to right) before being passed to the procedure, with the following six exceptions: **assigned**, **evaln**, **eval**, **evalhf**, **seq**, and **userinfo**. For the functions **assigned** and **evaln**, the arguments are evaluated to a name. For the functions **eval** and **evalhf**, no evaluation of arguments is performed prior to invoking the procedure, which then performs the specified type of evaluation. The function `seq` is evaluated using `for`-loop semantics, meaning that the index of sequence-formation (appearing in the second argument) is first evaluated to a name, and this name is assigned a specified value prior to each evaluation of the first argument. Finally, the function `userinfo` uses delayed evaluation of arguments beyond the first two in order to recognize occurrences of `print` and `lprint` for appropriate handling.

The Package Concept

The concept of a *package* in Maple permits the user to define an entire collection of new functions with just one Maple command, namely the `with` command. In addition to defining new user-level functions, packages provide two other services. First, they provide the necessary support functions to tie these routines properly into the rest of the system. This includes such facilities as type definitions and rules for expansion of mathematical expressions involving newly-defined functions. Secondly, they provide convenient short names for each of the routines in the package. In some cases, the effect is to redefine a name which is already used for some other purpose in the Maple session (or a name which is pre-defined by Maple). A warning is issued whenever such a redefinition of a name occurs via the `with` command.

Maple packages load very quickly because, in effect, loading a package requires nothing more than defining the locations of the new functions in the library. No code is loaded, but each of the member functions of a package becomes `readlib`-defined. For example, execution of the command

    with(linalg);

causes a table (named `linalg`) to be loaded. The indices of the table are the various function names, and the table entry for each index is the `readlib` definition of the corresponding code for the function. For example, `linalg[transpose]` has the value

    readlib(`linalg/transpose`)

Furthermore, the `with` command has the side-effect of assigning to each function name itself (such as the name `transpose`) the corresponding `readlib` definition. Hence, if the package is loaded via `with` then the simple function name `transpose` may be used; otherwise, the long-form notation `linalg[transpose]` would have to be used as the function name. As soon as the `transpose` function is first called, the code for this routine is loaded into memory.

12.3   Format of Library Function Descriptions

Each of the function descriptions in the on-line help system has the following sections:

- a one line description
- prototypes showing how the function is called
- information about what types of arguments are expected
- further detailed information about the function

- a few examples of the uses of the function
- a list of related functions or topics .

In the description of the function's parameters, the parameter names are generally prototypes, and may be replaced by expressions of the appropriate types according to the restrictions noted in the PARAMETERS section. Further conditions for the parameters may be described in the SYNOPSIS section. Other parameters are meant to appear verbatim. When such parameters exist, prototype parameters are generally enclosed in angle brackets ("<...>"). For example:

`int(f,x)`	`f` and `x` are prototypes.
`convert(<poly>,horner)`	`poly` is a prototype, `horner` should appear verbatim.

12.4   Printing Maple Help Files

Any of the descriptions in the on-line-help system can be saved in a text file and printed on your local printer. The following Maple procedure, ph, allows you to append help files to the file specified:

```
ph := proc(topic, file)
   local f, line;
   readlib(help):
   f := `help/help`(topic):
   if not type(f, specfunc(string,TEXT)) then
       ERROR(`could not find help file`)
   else
       if nargs = 2 then appendto(file) fi;
       for line in f do lprint(line) od;
       if nargs = 2 then appendto(terminal) fi;
   fi;
end;
```

Once the procedure **ph** has been read into your Maple session, the following set of commands creates a file, *outfile*, containing the help files for **int** and **diff**. (Note: If the help topic contains any special characters or keywords, it must be enclosed in backquotes.)

```
ph(int, outfile);
ph(diff, outfile);
```

12.5   Library Index

Reproduced below is a list of all the functions in the Maple library. A topic or the name of a function can be used as an argument to the ? command to get an on-line display of information about that topic or function.

`abs` . . . . absolute value of real or complex argument

`addressof`	obtain the address which points to an expression
`Ai, Bi`	Airy wave functions
`alias`	define an abbreviation or denotation
`allvalues`	evaluate all possible values of expressions involving RootOfs
`amortization`	amortization schedule
`anames`	sequence of assigned names
`appendto`	write output to a file in append mode
`array`	create an array
`assemble`	assemble a sequence of addresses into an object
`assign`	perform assignments
`assigned`	check if a name is assigned
`asympt`	asymptotic expansion
`bernoulli`	Bernoulli numbers and polynomials
`bernstein`	Bernstein polynomial approximating a function
`BesselI, BesselK`	Bessel functions
`BesselJ, BesselY`	Bessel functions
`Beta(x,y)`	the Beta function
`bianchi`	find the Bianchi type of any 3-dimensional Lie algebra
`binomial`	binomial coefficients
`blackscholes`	present value of a call option
`bspline`	compute the B-spline segment polynomials
`C`	generate C code
`cartan`	a collection of procedures for the computation of the connection coefficients and curvature components using Cartan's structure equations
`cat`	concatenating expressions
`chebyshev`	Chebyshev series expansion
`chrem`	Chinese Remainder Algorithm
`Ci`	the cosine integral
`close`	file I/O routines
`coeff`	extract a coefficient of a polynomial
`coeffs`	extract all coefficients of a multivariate polynomial
`coeftayl`	coefficient of (multivariate) expression
`collect`	collect coefficients of like powers
`combinat`	introduction to the `combinat` package
`combinat[bell]`	compute Bell numbers
`combinat[cartprod]`	iterate over a list of lists
`combinat[character]`	compute character table for a symmetric group
`combinat[Chi]`	compute Chi function for partitions of symmetric group
`combinat[combine]`	construct the combinations of a list
`combinat[composition]`	k-compositions of an integer
`combinat[decodepart]`	compute canonical partition represented by integer
`combinat[encodepart]`	compute canonical integer representing partition
`combinat[fibonacci]`	compute Fibonacci numbers or polynomials
`combinat[firstpart]`	first partition in canonical partition sequence
`combinat[inttovec]`	vector referenced by integer in canonical ordering
`combinat[lastpart]`	last partition in canonical partition sequence

convert/hex .... convert to hexadecimal form
convert/horner .... convert a polynomial to Horner form
convert/hostfile .... convert Maple filename to host filename
convert/hypergeom .... convert summations to hypergeometrics
convert/list .... convert to a list
convert/listlist .... convert to a list of lists
convert/ln .... convert arctrig functions to logarithms
convert/matrix .... convert an array or a list of lists to a matrix
convert/metric .... convert to metric units
convert/mod2 .... convert expression to mod 2 form
convert/multiset .... convert to a multiset
convert/octal .... convert to octal form
convert/parfrac .... convert to partial fraction form
convert/polar .... convert to polar form
convert/polynom .... convert a series to polynomial form
convert/radians .... convert degrees to radians
convert/radical .... convert RootOf to radicals and I
convert/rational .... convert float to an approximate rational
convert/ratpoly .... convert series to a rational polynomial
convert/RootOf .... convert radicals and I to RootOf notation
convert/series .... convert to series
convert/set .... convert to a set
convert/sincos .... convert trig functions to sin, cos, sinh, cosh
convert/sqrfree .... convert to square-free form
convert/std .... convert to standard form for simplex manipulation
convert/stdle .... convert inequalities to type $\leq$
convert/string .... convert an expression to a string (name)
convert/tan .... convert trig function to tan
convert/trig .... convert all exponentials to trigonometric and hyperbolic trigonometric functions
convert/vector .... convert a list or an array to a Maple vector
copy .... create a duplicate array or table
cost .... operation evaluation count
D .... differential operator
debever .... a collection of procedures for the computation of the Newman-Penrose spin coefficients and curvature components using Debever's formalism
define .... define characteristics of an operator name
define/forall .... define a property for an operator
define/group .... define characteristics of a group operator
define/linear .... define characteristics of a linear operator
define/operator .... define an operator by a list of properties
degree .... degree of a polynomial
denom .... denominator of an expression
Det .... inert determinant
diff or Diff .... partial differentiation
difforms .... introduction to the **difforms** package
difforms[&^] .... wedge product

evaln .... evaluate to a name

evalr .... evaluate an *expr* using range arithmetic

example .... produce an example of a function or type

exp .... the exponential function

Expand .... inert expand function

expand .... expand an expression

extrema .... find relative extrema of an expression

Factor .... inert factor function

factor .... factor a multivariate polynomial

Factors .... inert factors function

factors .... factor a multivariate polynomial

FFT .... fast Fourier transform

finance .... amount, interest, payment, or periods

fixdiv .... compute the fixed divisor of a polynomial

fnormal .... floating-point normalization

forget .... remove an entry or entries from a remember table

fortran .... generate Fortran code

frac .... fractional part of a number

freeze, thaw .... replace an expression by a name

FresnelC .... Fresnel cosine integral

Fresnelf .... auxiliary Fresnel function

Fresnelg .... auxiliary Fresnel function

FresnelS .... Fresnel sine integral

frontend .... process general expression into a rational expression

fsolve .... solve using floating-point arithmetic

galois .... compute the Galois group of an irreducible polynomial

GAMMA .... the gamma function

gc .... garbage collection

Gcd .... inert gcd function

gcd .... greatest common divisor of polynomials

Gcdex .... inert gcdex function

gcdex .... extended Euclidean algorithm for polynomials

genpoly .... generate polynomial from integer n by Z-adic expansion

geom3d .... introduction to the geom3d package

geom3d[angle] .... find the angle between two lines or two planes or a line and a plane

geom3d[area] .... compute the area of a triangle or the surface of a sphere

geom3d[are_collinear] ..... test if three points are collinear

geom3d[are_concurrent] .... test if three lines are concurrent

geom3d[are_parallel] ...... test if two lines or two planes or a line and a plane are parallel to each other

geom3d[are_perpendicular].. test if two lines or two planes or a line and a plane are perpendicular to each other

geom3d[are_tangent] .... test if a plane (line) and a sphere or two spheres are tangent to each other

geom3d[center] .... find the center of a given sphere

geom3d[centroid] .... compute the centroid of a given tetrahedron or a given triangle or a list of points in the space

geom3d[coordinates] .... return the coordinates of a given point

geom3d[coplanar] .... test if four points or two lines are coplanar

```
geometry[conic] ............ find the conic going through five points
geometry[convexhull] ....... find convex hull enclosing the given points
geometry[coordinates] ..... compute the coordinates of a given point
geometry[detailf] ......... give floating-point information about points, circles, and lines.
geometry[diameter] ........ compute the diameter of a given set or list of points on a plane.
geometry[distance] ........ find the distance between two points or a point with a line.
geometry[ellipse] ......... define the ellipse
geometry[Eulercircle] ..... find the Euler circle of a given triangle
geometry[Eulerline] ....... find the Euler line of a given triangle
geometry[excircle] ........ find three excircles of a given triangle
geometry[find_angle] ...... find the angle between two lines or two circles
geometry[Gergonnepoint] ... find the Gergonne point of a given triangle
geometry[harmonic] ........ find a point which is harmonic conjugate to another point with respect to
                            two other points
geometry[incircle] ........ find the incircle of a given triangle
geometry[inter] ........... find the intersections between two lines or a line and a circle or two circles.
geometry[inversion] ....... find the inversion of a point, line, or circle with respect to a given circle.
geometry[is_equilateral] ... test if a given triangle is equilateral
geometry[is_right] ........ test if a given triangle is a right triangle
geometry[line] ............ define the lines
geometry[make_square] ..... construct squares
geometry[median] .......... find the median of a given triangle
geometry[midpoint] ........ find the midpoint of segment joining two points
geometry[Nagelpoint] ...... find the Nagel point of a given triangle
geometry[on_circle] ....... test if a point, a list or set of points are on a given circle.
geometry[on_line] ......... test if a point, a list or a set of points are on a given line.
geometry[onsegment] ....... find the point which divides the segment joining two given points by some
                            ratio.
geometry[orthocenter] ..... compute the orthocenter of a triangle or of a set or list of points in a plane.

geometry[parallel] ........ the line which goes through a given point and is parallel to a given line.
geometry[perpen_bisector] .. find the line through the midpoint of two given points and perpendicular
                            to the line joining them
geometry[perpendicular] ... find a line which goes through a given point and is perpendicular to a given
                            line
geometry[point] ........... define the points
geometry[polar_point] ..... determine the polar line of a given point with respect to a given conic or a
                            given circle
geometry[pole_line] ....... determine the pole of a given line with respect to a given conic or a given
                            circle.
geometry[powerpc] ......... power of a given point with respect to a given circle
geometry[projection] ...... find the projection of a given point on a given line
geometry[rad_axis] ........ find the radical axis of two given circles
geometry[rad_center] ...... find the radical center of three given circles
geometry[radius] .......... compute the radius of a given circle
geometry[randpoint] ....... find a random point on a line or a circle
geometry[reflect] ......... find the reflection of a given point with respect to a given line
```

```
history .................... maintain a history of all values computed
hypergeom ................. generalized hypergeometric function
icontent .................. integer content of a polynomial
ifactor ................... integer factorization
ifactors .................. integer factorization
iFFT ...................... inverse fast Fourier transform
igcd ...................... greatest common divisor of integers
igcdex .................... extended Euclidean algorithm for integers
ilcm ...................... least common multiple of integers
indets .................... find indeterminates of an expression
indices ................... indices of a table or array
int or Int ................ definite and indefinite integration
interface ................. set or query user interface variables
Interp .................... inert interp function
interp .................... polynomial interpolation
invlaplace ................ inverse Laplace transform
invztrans ................. inverse Z transform
iquo ...................... integer quotient
iratrecon ................. rational reconstruction
irem ...................... integer remainder
```

iroot .... integer n^{th} root

iroot .... integer n^{th} root

```
Irreduc ................... inert irreducibility function
irreduc ................... polynomial irreducibility test
iscont .................... test continuity on an interval
isolate ................... isolate a subexpression to left side of an equation
isolve .................... solve equations for integer solutions
isprime ................... primality test
isqrfree .................. integer square free factorization
isqrt ..................... integer square root
isqrt ..................... integer square root
issqr ..................... test if an integer is a perfect square
ithprime .................. determine the ith prime
laplace ................... Laplace transform
latex ..................... produce output suitable for latex printing
latex[functions] ......... how latex formats functions
latex[names] ............. how latex formats names
lattice ................... find a reduced basis of a lattice
lcm ....................... least common multiple of polynomials
lcoeff .................... leading coefficient of a multivariate polynomial
ldegree ................... low degree of a polynomial
length .................... length of an object
lexorder .................. test for lexicographical order
lhs ....................... left hand side of an expression
liesymm ................... introduction to the liesymm package
liesymm[annul] ........... annul a set of differential forms
```

linalg[delcols] .... delete columns of a matrix
linalg[delrows] .... delete rows of a matrix
linalg[det] .... determinant of a matrix
linalg[diag] .... create a block diagonal matrix
linalg[diverge] .... compute the divergence of a vector function
linalg[dotprod] .... vector dot (scalar) product
linalg[eigenvals] .... compute the eigenvalues of a matrix
linalg[eigenvects] .... find the eigenvectors of a matrix
linalg[equal] .... determine whether two matrices are equal
linalg[exponential] .... matrix exponential
linalg[extend] .... enlarge a matrix
linalg[ffgausselim] .... fraction-free Gaussian elimination on a matrix
linalg[fibonacci] .... fibonacci matrix
linalg[frobenius] .... compute the Frobenius form of a matrix
linalg[gausselim] .... Gaussian elimination on a matrix
linalg[gaussjord] .... a synonym for rref (Gauss-Jordan elimination)
linalg[genmatrix] .... generate the coefficient matrix from equations
linalg[grad] .... vector gradient of an expression
linalg[GramSchmidt] .... compute orthogonal vectors
linalg[hadamard] .... bound on coefficients of the determinant of a matrix
linalg[hermite] .... Hermite normal form (reduced row echelon form)
linalg[hessian] .... compute the Hessian matrix of an expression
linalg[hilbert] .... create a Hilbert matrix
linalg[htranspose] .... compute the Hermitian transpose of a matrix
linalg[ihermite] .... integer-only Hermite normal form
linalg[indexfunc] .... determine the indexing function of an array
linalg[innerprod] .... calculate the inner product
linalg[intbasis] .... determine a basis for the intersection of spaces
linalg[inverse] .... compute the inverse of a matrix
linalg[ismith] .... integer-only Smith normal form
linalg[iszero] .... determine whether a matrix is zero
linalg[jacobian] .... compute the Jacobian matrix of a vector function
linalg[jordan] .... compute the Jordan form of a matrix
linalg[JordanBlock] .... return a Jordan block matrix
linalg[kernel] .... compute a basis for the null space
linalg[laplacian] .... compute the Laplacian
linalg[leastsqrs] .... least-squares solution of equations
linalg[linsolve] .... solution of linear equations
linalg[matrix] .... create a matrix
linalg[minor] .... compute a minor of a matrix
linalg[minpoly] .... compute the minimum polynomial of a matrix
linalg[mulcol] .... multiply a column of a matrix by an expression
linalg[mulrow] .... multiply a row of a matrix by an expression
linalg[multiply] .... matrix-matrix or matrix-vector multiplication
linalg[norm] .... norm of a matrix or vector
linalg[nullspace] .... a synonym for kernel

maxnorm	infinity norm of a polynomial
MeijerG	special case of the general Meijer G function
mellin	Mellin transform
member	test for membership in a set or list
min	minimum of numbers
minimize	compute the minimum
minpoly	find minimum polynomial with an approximate root
&mod	reduce a form modulo an exterior ideal
mod, modp, mods	computation over the integers modulo m
modp1	univariate polynomial arithmetic modulo n
modpol	expression evaluation in a quotient field
MOLS	mutually orthogonal Latin squares
msolve	solve equations in Z mod m
mtaylor	multivariate Taylor series expansion
Newman-Penrose commutators	V_D, X_D, Y_D, X_V, Y_V, Y_X
Newman-Penrose Pfaffians	D, V, X, Y
nextprime	determine the next largest prime
nops	the number of operands of an expression
norm	norm of a polynomial
Normal	inert normal function
normal	normalize a rational expression
np	introduction to the **np** package
np[conj]	Newman-Penrose complex conjugation operator
np[eqns]	initialization and display of NP equations
np[suball]	substitution utility for Newman-Penrose package
Nullspace	compute the nullspace of a matrix mod p
numer	numerator of an expression
numtheory	introduction to the **numtheory** package
numtheory[cfrac]	continued fraction convergents
numtheory[cyclotomic]	calculate cyclotomic polynomial
numtheory[divisors]	positive divisors of an integer
numtheory[factorset]	prime factors of an integer
numtheory[fermat]	nth Fermat number
numtheory[GIgcd]	gcd of Gaussian integers
numtheory[imagunit]	square root of -1 mod n
numtheory[issqrfree]	test if integer is square free
numtheory[jacobi]	Jacobi symbol
numtheory[lambda]	Carmichael's lambda function
numtheory[legendre]	Legendre symbol
numtheory[mcombine]	Chinese remaindering
numtheory[mersenne]	nth Mersenne prime
numtheory[mipolys]	number of monic irreducible univariate polynomials
numtheory[mlog]	discrete logarithm
numtheory[mobius]	Mobius function
numtheory[mroot]	modular root
numtheory[msqrt]	modular square root
numtheory[nthpow]	find largest nth power in a number

`RETURN`	explicit return from a procedure
`rhs`	right hand side of an expression
`RootOf`	a representation for roots of equations
`Roots`	roots of a polynomial mod n
`roots`	roots of a univariate polynomial
`round`	round a number to an integer
`rsolve`	recurrence equation solver
`search`	substring search
`select`	selection from a list, set, sum, or product
`seq`	create a sequence
`series`	generalized series expansion
`series/leadterm`	find the leading term of a series expansion
`shake`	compute a bounding interval
`showprofile`	space & time profile of a procedure
`showtime`	display time and space statistics for commands
`Si`	the sine integral
`sign`	sign of a number or a polynomial
`signum`	sign function for real and complex expressions
`simplex`	introduction to the `simplex` package
`simplex[basis]`	computes a list of variables, corresponds to the basis
`simplex[convexhull]`	finds convex hull enclosing the given points
`simplex[cterm]`	computes the list of constants from the system
`simplex[dual]`	computes the dual of a linear program
`simplex[feasible]`	determine if system is feasible or not
`simplex[maximize]`	maximize a linear program
`simplex[minimize]`	minimize a linear program
`simplex[pivot]`	construct a new set of equations given a pivot
`simplex[pivoteqn]`	returns a sublist of equations given a pivot
`simplex[pivotvar]`	returns a variable with positive coefficient
`simplex[ratio]`	returns a list of ratios
`simplex[setup]`	constructs a set of equations with variables on the lhs
`simplex[standardize]`	converts a set of equations to type $\leq$
`simplify`	apply simplification rules to an expression
`simplify/atsign`	simplify expressions involving operators
`simplify/GAMMA`	simplifications involving the GAMMA function
`simplify/hypergeom`	simplify hypergeometric expressions
`simplify/power`	simplify powers
`simplify/radical`	simplify expressions with radicals
`simplify/RootOf`	simplify expressions with the RootOf function
`simplify/siderels`	simplify with respect to side relations
`simplify/sqrt`	simplify square roots
`simplify/trig`	simplify trigonometric expressions
`singular`	find singularities of an expression
`sinterp`	sparse multivariate modular polynomial interpolation
`Smith`	compute the Smith normal form of a matrix mod p
`solve`	solve equations
`solve/floats`	expressions involving floating-point numbers

stats[statplot] .... statistical plotting
stats[StudentsT] .... the Student's T distribution
stats[Uniform] .... the Uniform distribution
stats[variance] .... compute the variance of a set of data
student .... introduction to the **student** package
student[changevar] .... perform a change of variables
student[combine] .... combine terms into a single term
student[completesquare] ... complete the square
student[distance] .... compute the distance between **Points**
student[Int] .... inert form of int (integration function)
student[intercept] .... compute the points of intersection of two curves
student[intparts] .... perform integration by parts
student[isolate] .... isolate a subexpression to left side of an equation
student[leftbox] .... graph an approximation to an integral
student[leftsum] .... numerical approximation to an integral
student[Limit] .... inert form of limit
student[makeproc] .... convert an expression into a Maple procedure
student[maximize] .... compute the maximum
student[middlebox] .... graph an approximation to an integral
student[middlesum] .... numerical approximation to an integral
student[midpoint] .... compute the midpoint of a line segment
student[minimize] .... compute the minimum
student[powsubs] .... substitute for factors of an expression
student[rightbox] .... graph an approximation to an integral
student[rightsum] .... numerical approximation to an integral
student[showtangent] ...... plot a function and its tangent line
student[simpson] .... numerical approximation to an integral
student[slope] .... compute the slope of a line
student[Sum] .... inert form of sum
student[trapezoid] .... numerical approximation to an integral
student[value] .... evaluate inert functions (formerly student[Eval])
sturm .... number of real roots of a polynomial in an interval
sturmseq .... Sturm sequence of a polynomial
subs .... substitute subexpressions into an expression
subsop .... substitute for specified operands in an expression
substring .... extract a substring from a string
Sum .... inert form of summation
sum .... definite and indefinite summation
Svd .... compute the singular values/vectors of a numeric matrix
system .... invoke a command in the host operating system
table .... create a table
taylor .... Taylor series expansion
tcoeff .... trailing coefficient of a multivariate polynomial
tensor .... compute curvature tensors in a coordinate basis
testeq .... random polynomial-time equivalence tester
thiele .... Thiele's continued fraction interpolation formula

type/radical	check for fractional powers
type/radnum	check for an algebraic number in terms of radicals
type/radnumext	check for a radical number extension
type/range	check for a range
type/rational	check for an object of type rational
type/ratpoly	check for a rational polynomial
type/realcons	check for a real constant
type/RootOf	check for a RootOf expression
type/scalar	check for scalar (in the matrix sense)
type/scalar	check for scalars
type/sqrt	check for a square root
type/square	check for a perfect square
type/taylor	check for Taylor series
type/trig	check for trigonometric functions
type/type	check for type expressions
type/vector	check for vector (one-dimensional array)
unames	sequence of unassigned names
unapply	returns an operator from an expression and arguments
unassign	unassign names
unprofile	space & time profile of a procedure
userinfo	print useful information to the user
W	the omega function
whattype	query the basic data type of an expression
with	define the names of functions from a library package
words	query memory usage (words used)
write	file I/O routines
writeln	file I/O routines
writeto	write output to a file
Zeta	the Riemann zeta function
zip	zip together two lists or vectors
ztrans	Z transform
&^	the wedge product

Appendix A
Maple under UNIX

A.1 Introduction

Maple is not designed to provide a completely self-contained programming environment; rather, it is designed to work with other services provided by the operating system of the computer on which it runs. Using Maple under the UNIX operating system requires some knowledge of the fundamentals of UNIX usage, such as logging in and out, using a text editor, and listing and deleting files. This appendix assumes that the reader knows how to do such activities without providing further explanation. Such explanations can be found in any basic text on UNIX.

The standard installation instructions for Maple make it possible for UNIX users to start a Maple session by typing `maple`. The related program Mint should also be available as a system command `mint`. Information is available on-line via the UNIX `man` command; for example, `man maple`.

A.2 Maple Initialization Files

The file `src/init` in the Maple library plays a special role for all Maple sessions under UNIX. If that file exists, it is read (in the sense of the Maple **read** statement) as the first action in any Maple session, before any Maple commands from the user are executed. The `init` file typically contains Maple commands establishing start-up values for Maple system variables, such as those controlled by `gc` and `words`. It is typically set up by the system manager to establish start-up values for such variables that are appropriate for the site.

If a user has a file named `.mapleinit` in his or her home directory, that file is read into Maple after the system-wide `src/init` file. The user can put Maple commands into `.mapleinit` that supplement or override those found in the `src/init` file. However, many users will find that they don't need a `.mapleinit` file if the default start-up actions work satisfactorily for them.

A.3   Quit and Interrupt Characters

Maple uses the user's UNIX quit (for many users, *Control-*) and interrupt (for many users, *delete* or *break*) signals to interrupt a Maple calculation. If the quit character is typed, the Maple session will be aborted, and control returns to the user's shell. If the interrupt character is typed once, the currently executing Maple calculation will be aborted, and another Maple command can then be entered. If the interrupt character is typed twice within three seconds, the Maple session will be aborted as though an interrupt character had been typed. On Berkeley UNIX systems, the "suspend" character (typically, *Control-Z*) works with Maple as with other foreground processes.

A.4   Temporarily Escaping from Maple

When interacting with Maple, typing a line beginning with ! means that what follows is to be interpreted and executed as a UNIX command instead of a Maple statement or command. All UNIX commands, such as !vi or !who, are accessible via this mechanism. When execution of the UNIX command is finished, the Maple session is resumed where it had been suspended. The results of the command do not affect the state of the Maple session when it is resumed.

A.5   Redirection of Input and Output

Maple on UNIX uses the standard input and output files for reading and printing information. By default, these direct input from and output to the user's terminal. One can change this by employing the usual means of the UNIX command language. Invoking the `maple` command with a < and a file name results in the specified file being used as the input source of a Maple session instead of the terminal keyboard. Once all lines have been read from the file, additional input may be entered through the terminal keyboard (unless the session has ended because the file contained a Maple **quit**, **done**, or **stop** command).

Similarly, if Maple is invoked with > and a file name, then the results of the Maple session will be placed in the specified file instead of the terminal. Input redirection and output redirection may be used together.

Example:

```
maple < file
```

will use the contents of `file` as the input to the Maple session instead of the terminal keyboard.

```
maple < input_file > output_file
```

will start up a Maple session, take input from `input_file`, and place the results of the Maple session in `output_file`. Note that `output_file` will not contain the commands (from `input_file`), but only the results of commands.

A.6 Maple Command Line Options for UNIX

A.6.1 Overview of Maple command line options

Maple can be invoked with up to three options on the command line: `-s`, `-q`, and `-b`. The effect and purpose of each of them is described in the following section. Users will not need to include any of them on the command line for typical interactive usage of Maple.

Command line options can be used together with redirection of input and output from and to files, but then the command line options must appear before the < or >.

Example:

```
maple -q < input_file > output_file
```

invokes Maple with the `-q` option, taking input from `input_file` and placing results in `output_file`.

A.6.2 Library Specification Option: -b

The Maple variable `libname` names the default location of the Maple library directory. The command `maple -b` *location* invokes Maple where this default is overridden, using *location* as the location (directory) of the Maple library instead. *location* replaces the default value of the Maple variable `libname` for that session.

A.6.3 Suppress Initialization Option: -s

`maple -s` will cause Maple to forego reading any initialization files (see Section A.2).

A.6.4 Quiet Option: -q

`maple -q` will suppress the printing of Maple's startup logo and various informational messages ('`bytes used`' messages and garbage collection messages).

A.7   Mint

Mint is a program which checks a Maple source file for syntax errors and gives warnings about possible problems. Additionally, it can report how variables are used within procedures defined in a Maple source file. The general functionality of the Mint program is explained in Chapter 11 (see Section 11.1).

The most common syntax for the Mint command on UNIX is:

        mint *input_file*

which invokes Mint to process the Maple procedures and commands contained in the Maple source file *input_file*. More generally, the amount of information reported by the Mint command is controlled by an *information level* flag as explained in Chapter 11. The syntax for specifying this flag is:

        mint -i *info_level input_file*

where *info_level* is an integer between zero and four.

There are additional optional arguments to the Mint command, with the general format:

        mint ‖ -i *info_level* ‖
                   ‖ -l   ‖
                   ‖ -d *library_database* ‖
                   ‖ -a *database_file* ‖
                   ‖ -q   ‖
                   ‖ *input_file* ‖

The ‖ 's indicate optional arguments. See man mint on UNIX for an explanation of the optional arguments.

A.8   Summary of Site- and UNIX- Dependent Aspects of Maple

Site dependencies

Invoking Maple

Type maple after seeing the UNIX prompt. Check with your local system consultants if this does not work.

Owner of Maple library and code

Ask your system manager for information about your site, or use the `ls -l` command on the Maple library directory.

Location of Maple library

Given by Maple variable `libname`

Location of Maple system code

Ask your system manager for information about your site.

Access to Maple code

Typically, users have read access to Maple library files and execute access to `maple` and `mint`.

UNIX Dependencies

Quit signal

Will exit Maple immediately.

Interrupt signal
(*delete* **or** *break*)

Abort currently executing Maple command, but remain in Maple.

Escape to UNIX

! *command* escapes Maple temporarily to perform *command*. After it finishes executing, control returns to the Maple session. (See also the Maple built-in function `system`.)

Maple file names

Maple file names are the same style as UNIX file names. The Maple command `convert` (*filename*, `hostfile`) on UNIX systems returns its *filename* argument without transformation.

Session initialization

Defined by the system manager in the Maple library file `src/init`.

User's own initialization

`.mapleinit` in the user's home directory.

The `words` function

Words of memory are measured in units of 32 bits (four bytes) and reported in `bytes used` messages.

The `time` function

`time` reflects CPU cycles devoted to Maple execution, measured in seconds, as reflected by the `time=` messages.

Character set

All ASCII characters (including control characters, et cetera).

Mint

```
mint   input_file
mint -i  info_level input_file
mint || -i  info_level ||
          || -l  ||
          || -d  library_database ||
          || -q  ||
          || -a  database_file ||
          || input_file ||
```

Help Files, `helptomaple`

Convert user help files to Maple TEXT format. Usage,

```
helptomaple helpname < myhelpfile > outfile
```

where `myhelpfile` is the name of an user text help file, and `helpname` is the name you would like to use in Maple to bring up the help page, and `outfile` is the file to be read in by Maple. For example, to be able to bring up a help page for your function `foo` in Maple, you should create a text file `foo.help` to describe the function, then execute the UNIX command,

   `helptomaple < foo.help > foo.M`

After executing the following command in Maple,

   ``read `foo.M`;``

you should be able to bring up the help for `foo` with the command `?foo`.

Appendix B
Using Maple with X

B.1 Introduction

This appendix describes the user interface features for the X windows version of Maple. This version of Maple is typically invoked by typing `xmaple` and is designed to run in the X windowing environment. Some of the important features of this version of Maple are

- A graphics display driver for two and three-dimensional plotting. Each plot appears in its own window complete with menu options, for example for interactively rotating a three-dimensional graph.

- A Maple session window with editing capabilities, a search facility, and utilities for file inclusion and session saving.

- The on-line documentation appears in seperate windows. Examples may easily be cut and pasted into the session window.

B.2 Getting Started

The first thing that you will notice as you invoke Maple with the X Window System interface is that a Maple worksheet window is created on the screen. This window will display the Maple leaf logo in its top left corner and will contain a row of four buttons across the top. It is within this window that you will enter your Maple commands and see the results of the computations. In addition, other windows with on-line documentation and graphics may also be created during your use of Maple. The exact appearance of these windows may differ from system to system and is affected by both the window manager that you use and the X resource settings for Maple.

If you do not see a new window being created, then you probably have invoked Maple without specifying that it should run along with its own X Window System user interface program. An indication that this has happened is the display of the Maple leaf logo within your xterm or other terminal emulation window. If this occurs, please consult the installation instructions for Maple or check with your local systems consultant for the proper command to use. Most systems will provide the command **xmaple** to run Maple using its X Window user interface and the command **maple** to run Maple as a program using an ASCII text terminal.

If you see messages indicating that Maple cannot open a window on your display screen, then you should consult Section B.16: Troubleshooting.

There are many window managers available (awm, twm, uwm, mwm, dxwm, etc.) and we assume that you are using one such program. The window manager will determine the way in which the window frame, the title bar, and the vertical scroll bar are drawn. Each window manager will provide methods for you to raise a window above other windows on the screen, to change the size of the window, and to move the window to different locations on the screen. Since each window manager provides different methods for these procedures, we won't comment on them any further and we assume that you are already familiar with the necessary techniques. Proficient use of Maple under X Window requires you to be able to move easily between the worksheet window, on-line documentation windows, and graphics windows. Fortunately, even if you are not an experienced user of the X Window System, these techniques can be mastered with very little practice.

To work within a window you must first make it active. If you have multiple windows on the screen, only one of them will be active at any given time. Clues that window managers provide in identifying the active window are a different border color or a different appearance for the title bar of the active window. With some window managers, simply moving the cursor on the screen from window to window by moving the mouse is enough to change the active window. With these window managers, the active window is always the window in which the cursor is located. With other window managers, you may have to click the cursor within the appropriate window to activate it. You will find it convenient to have the active window raised above all other windows. If the Maple worksheet window is not raised automatically by your window manager when it becomes the active window, then you should raise the window. Only the active window will recognize any input which you may type.

The behavior and appearance of windows, menus, and buttons under the X Window System can be easily customized through various means. To simplify the explanations in the following sections, we will assume that the default bindings and default resource settings are used. If the keyboard keys and mouse buttons don't work as described or if the appearance of buttons is not as described, then consult Section B.14: Customizing Maple Under X.

B.3   Entering Commands in Maple

Once the Maple worksheet window is active, you can begin typing Maple commands. You should notice that there is a prompt character (normally **>**) at left side of the window. The Maple prompt appears to let you know that Maple is now ready to receive more input, usually the next Maple command. Also notice that there is a small caret (ˆ) character appearing after the prompt. Any characters that you type will be inserted into the window at the location of the input caret. Try typing a simple Maple command such as **10!;** to compute 10 factorial.

Once you have typed any single-line Maple command, you can press the return key to have Maple evaluate the expression and produce an answer. Usually single-line commands suffice and this method of entering and evaluating commands is both easy and natural. If the expression that you type is large, you may need to press the return key to continue typing on the next line in the window. When you do this, the first line is read and parsed by Maple and, assuming that the command has not been completely typed, Maple will display another prompt for the rest of the command. When you have finished typing all of the command, perhaps using several more lines, ending with the normal command terminator(;), press the return key and Maple will evaluate the complete expression and produce an answer.

An alternative for entering multiple line Maple commands or expressions line by line is to use the newline or enter key. (If there is no separate newline or enter key on your keyboard, the control-J key combination can be used.) As long as you type and then use the newline key, you can move to the next line in the window for further typing without having Maple read what you have typed so far and parse it as part of a command. This has the advantage of giving you the ability to type a multi-line expression and then to review the entire expression for errors. Techniques given below describe how you can edit this expression within the window. Finally, you can select the entire multi-line expression for evaluation by Maple. To do this, using the mouse, move the cursor to the point just left of the the first prompt character for the expression, click on the left mouse button, and while holding the left button down, drag the mouse down over the entire expression. Now release the button. What you should see is the entire expression highlighted. We will call this method of selecting a block of text *drag-selection*. After using drag-selection to highlight the entire expression, you may now press the return key to have it read and evaluated by Maple.

Using the newline key to type information within the window without having it read as a Maple command is useful for a different purpose. It allows you to annotate the worksheet with comments. In fact, you can add anything that you wish from simple one-line comments to many paragraphs of explanations onto the worksheet. Just remember to use the newline key instead of the return key at the end of a line. If you are typing many lines of documentation and want to switch the roles of the newline key and the return key, you can toggle the menu item **Command Key: Return** in the **Utilities** menu. If there is no newline or enter key on your keyboard and you want a different key or key combination to serve this purpose, consult Section B.14: Customizing Maple Under X.

Text which appears anywhere within the Maple worksheet window may be used as a Maple command. You can move the caret text insertion character to a different location by moving the mouse, and clicking the left mouse button to relocate the cursor. Pressing the return key immediately afterwards passes to Maple the entire line of text in which the caret is positioned. The caret character can be positioned anywhere on the line for the entire line to be read. When the return key is pressed just after a block of text has been highlighted through drag-selection, the entire selection is passed to Maple for reading and evaluation. The selection may cover multiple lines or it may be just part of one line.

B.4   Editing

Text which is typed within a window can be edited very easily. The delete, erase, or backspace key erases the character just to the left of the caret character. Which of these keys is appropriate for this depends on the keyboard that you use and your Unix terminal settings. (See the documentation for the Unix command stty.) You can erase characters anywhere within the window. Just relocate the caret character by moving the cursor and clickng on the left mouse button at the appropriate spot before using the erase key. An entire block of text, even spanning multiple lines, can be quickly deleted by first drag-selecting the text and then by pressing the control-W key combination.

Characters can be inserted easily into any line in the window simply by changing the position of the caret character before typing.

Characters can also be inserted into the window through copy and paste operations. Any text that has been selected is automatically copied into an internal buffer maintained by the X Window System. In the previous section on entering commands in Maple, you were introduced to one method of selecting text, drag-selection. Text copied into the internal buffer through text selection can be pasted into arbitrary locations in the window. To insert text from the buffer at a certain position, move the cursor to that position and click on the left mouse button. This will reposition the caret character to the appropriate location. Pressing the middle mouse button will insert the text from the buffer.

There are other methods of selecting text and hence copying the text into the buffer. One method is to click at the beginning of block of text using the left mouse button, release the button, move the cursor to the end of the block of text, and click using the right mouse button. This selects and highlights the entire block of text between the two positions that were chosen. A method of selecting just one word is to position the cursor anywhere on the word and then click on the left mouse button twice quickly. This double-click operation should select and highlight the entire word. Triple-clicking is an easy method for selecting an entire line.

You will find these copy and paste operations useful for editing comments inside your worksheet, for editing and then re-evaluating Maple commands, and also for copying parts of Maple output for re-use in further Maple commands.

B.5   Maple Input and Output Cells

As you enter Maple commands and as Maple produces the corresponding output results, you will notice that each input and output pair are separated into cells by separator lines which are drawn across the window. These lines are a visual aid in distinguishing between groups of input and output expressions. The appearance of these lines can be enabled or disabled through the use of the menu item **Separator Lines** in the **Utilities** menu. The menu item **Remove Separators** erases any separator lines which are present in the worksheet.

It is often useful to edit a previous Maple command and then enter it for re-evaluation. When this happens, you will notice that any former output results shown after the command are erased and then replaced by the new output. This maintains the worksheet in an orderly fashion by removing older results that may no longer be useful. Each cell will continue to contain one input expression and one output expression even after the input expression is recomputed. If you would like to keep all output results, then the menu item **Replace Mode** in the **Utilities** menu can be toggled off. A check mark shown alongside that menu item indicates that replacement of output is currently in effect.

When the replace mode is used, after a Maple command has been entered, Maple will erase all lines after the command until it reaches a line that it determines cannot be part of an output result. This includes a line beginning with the Maple prompt, a line beginning with the Maple comment character (#), or a output separator line.

B.6   Including and Saving Text

It is sometimes useful to include text from a file into the Maple worksheet. The file may contain documentation, Maple commands, or any other type of text input. You can include the contents of a text file by selecting the menu item **Include File** within the **Utilities** menu. This produces a dialog box on the screen that prompts you for the name of a file. Move the cursor within the rectangular text entry box at the top of the dialog box and type the name of the file to be included. Finally, confirm your choice by clicking on the **OK** button or by pressing the return key. You may also click on the **Cancel** button if you don't wish to proceed any further.

All the text which is contained in your worksheet window may be saved into a file. To do this, just select the **Save File** menu item on the **Utilities** menu. This produces a dialog box similar to the one for **Include File** to let you specify the name of a file into which the text from the worksheet window will be written.

B.7   Searching

When you are working with a large document or worksheet, it is often useful to be able to search for certain variables or pieces of text. A textual search utility is provided as an aid for this. By pressing the control-S key combination, you will produce a dialog box that prompts you for a search string. By pressing the **Search** button in the dialog box, you will initiate a search in the active text window, whether a worksheet window or an on-line documentation window, for the string that you specified. The search begins in the forward direction from the present selection location or the location of the caret character in the text window. By pressing the **Backward** direction button, you can indicate that searching is to proceed backwards from the current location.

The search dialog box also allows you to type a replacement string that can be used to replace an occurrence of the search string. The buttons **Replace** and **Replace All** will replace the next occurrence or all occurrences in the direction of the search respectively, of the search string by the replacement string.

The key combination control-R has the same effect as control-S except that it initiates the search in the reverse or backwards direction.

B.8   Resource Usage

At the bottom of the Maple worksheet window, two values are displayed. The value labelled **Bytes Allocated** shows the amount of memory that has been allocated for the main Maple program and for the data structures that it uses. The other value, labelled **CPU Time**, shows the cumulative CPU time in seconds used by the main Maple program. Both these meters give you an indication of the amount of space and time used by Maple as it is working. These values are updated periodically while Maple is working on a computation.

B.9   Interrupt, Pause and Quit Buttons

While Maple is working on a computation, it will change the cursor from the normal I-beam cursor to a watch cursor. This indicates to you that the program is in the middle of an operation that may take some time. You can watch the resource usage meters at the bottom of the window to see how hard Maple is working on a particular computation. If you should decide to abort the current computation, you can move the cursor to the **Interrupt** button at the top of the window and then press the left mouse button. We call this operation "pressing the **Interrupt** button". This should terminate the current computation and Maple should then prompt you for further commands in the worksheet.

Pressing the **Pause** button will temporarily suspend the generation of a large output result. Pressing the **Pause** button again will resume the generation of the output. You can use the worksheet window's scroll bar to move up and down within an output result that is larger than the length of the window.

Finally, the **Quit** button may be pressed to terminate the Maple program. This is an alternative to typing the Maple command `quit`. When the Maple program quits, it cleans up by disposing any of the on-line documentation windows and any graphics windows that it may have created.

B.10   Resizing Windows

As more windows share the space on your display screen, you may find it useful to change the size of some windows as well as to reposition them within the screen. For example, you may find it helpful to have Maple display an on-line documentation window next to your worksheet window. For this, it may be appropriate to resize the window so that it is narrower than usual. The correct technique for resizing a window depends on the window manager that you use.

When a Maple worksheet window is resized, subsequent output expressions will be formatted by Maple to fit across the new width of the window. This occurs automatically as the window is resized. Previous output expressions already displayed will not be reformatted. Therefore if you change the window so that it is narrower, parts of former output expressions may no longer appear within the window. They are not lost and can be viewed in their entirety by widening the window again.

B.11   Help Windows

Within the Maple system, you can ask for on-line documentation to be shown for many topics. The command ?*topic* will produce any documentation that is available on *topic*. The command ? by itself will explain how to use the on-line help system. Under the X Window System, Maple on-line documentation appears in a separate help window. The text searching facility described for the worksheet window can also be used for a help window. Text and examples from the help window can be copied into X's internal buffer through any of the text selection techniques. These can then be copied into the worksheet for editing and further use.

A good technique for learning about some of the functions available in Maple is to first ask for the on-line documentation for a function, select one of the examples of the uses of the function which appear at the end of the documentation, and paste it back into the worksheet. You can then evaluate it immediately from the worksheet or you can edit the example to change some of the parameters before evaluating it.

The **Close** button at the top left corner of the help window removes the window when you are finished with it.

B.12   2D Plot Windows

The 2D plot windows created by Maple's `plot` command are very simple. The top left corner contains a **Close** button that you may press to remove the plot window after you are finished reviewing the graph. There is a pane at the bottom of the window that shows the x and y coordinates of points in the plot. Each time that you move the cursor about within the area of the plot and then press one of the mouse buttons, the location of the cursor will be shown in x and y coordinates.

B.13   3D Plot Windows

The 3D plot windows created by Maple's `plot3d` command display a graph according to the style and view options set either explicitly in the `plot3d` call or implicitly with default values. After the graph has been displayed you can actively manipulate it and change any of the many style and view options. A row of buttons along the top of the window allow you to change many of the option settings. These buttons are **Close**, **Projection**, **Plot Style**, **Scale Type**, **Color**, **Axes Style**, **Theta**, **Phi**, **Print**, and **Plot**. In addition, there is a horizontal scrollbar under the buttons and above the graph which can be used to control the amount of perspective transformation. Finally, one further control which you can use to change the display options is to place the cursor on top of the graph itself, press down on the left mouse button, and, while keeping the mouse button depressed, drag the surface of the plot into a new viewpoint orientation. After changing any of the display options, press the **Plot** button to have the graph redrawn.

The **Projection** button brings up a menu of two choices for setting the amount of perspective transformation used: **Orthogonal** and **Perspective**. These corresponds to the `plot3d` options of `projection=1` and `projection=0` respectively. In addition, you can grab the box inside the horizontal scrollbar for the perspective setting and drag holding down the left mouse button, to set a value for projection anywhere between 0 (on the left) and 1 (on the right).

The **Plot Style** button allows you select from the four different plot display styles: **Wire Frame**, **Hidden Line Removal**, **Color Patch**, and **Plot Points Only**. These are explained in Chapter 10.

The **Scale Type** button lets you choose between the options **Constrained**, which forces `plot3d` to respect a 1:1:1 aspect ratio between the scaling of the axes, and **Unconstrained**, which allows `plot3d` to stretch the plot to fit the plot window.

The **Color** button gives you a choice between the **XYZ**, **XY**, and **Z** coloring schemes for coloring the graph. This button may be dimmed and inactive if the device that you use does not support different colors. As more plot windows are displayed simultaneously, more colors are used in the display's color table. At some point, you may notice that subsequent plot windows don't use color in drawing their graphs since there aren't sufficient slots left in the device's color table.

The **Axes Style** button lets you select from these drawing styles for the axes: **XYZ Frame**, **Boxed**, **No Axes,** and **Normal Axes**. Each of these styles is explained in Chapter 10.

The **Theta** and **Phi** buttons display the current angles (in degrees) for the theta and phi orientation view angles. Pressing either of these buttons displays a dialog box that allow you enter a specific number for a view angle. In addition, you can use the left mouse button to drag the surface of the graph back and forth to change the viewing angle. As the surface is being rotated, you will see a bounding box being continually redrawn to show you the new orientation. As is true for changing any of the display settings, you need to press the **Plot** button to have the graph redisplayed using the new settings.

The **Print** button brings up a dialog box that lets you save the plot into a file. The graphics formats available are pic or PostScript. Buttons in the dialog box lets you select either one of these formats. Inside the dialog box there is a text box that allows you to enter the name of the plot output file. Initially it will be `plotoutfile`, the default output file.

Finally, the **Close** button can be used to remove the graphics window when you are finished viewing the graph.

B.14   Customizing Maple Under X

Typical programs that make use of the X Window System rely upon resource files to set or change the appearance of many of the items within the windows. Maple follows this model and uses a resource file named Maple that is distributed along with the Maple program. If the Maple program running under the X Window system cannot locate this file on your system, then the Maple window will look very odd. For example, the label on the **Quit** button will be **quitbutton** rather than **Quit**. If you notice this, then you should either consult the installation instructions that came with Maple or a local systems consultant. Normally, the Maple resource file would be installed in the system-wide application defaults directory **/usr/lib/X11/app-defaults**.

One resource setting that is often changed is the font that is used in the Maple worksheet window. Different users may want to choose a larger or a smaller font according to the size of their display unit, screen resolution, or the need to see more information at one time. You can change the font through the addition of a line such as:

```
Maple*Font:            12x20
```

to a resource file read by your X server program. This line declares that the Maple window should use the font named **12x20**. This line should be added to a file such as **.Xdefaults** in your home directory. The lines

```
maplehelp*font:        12x20
maple2dplot*font:      12x20
Maple3D*font:          12x20
```

set the font to be used in the Maple help, Maple 2D plot, and Maple 3D plot windows, respectively, to the **12x20** font. The program **xlsfonts** will produce a list of fonts that are available on your system. For best results, choose a mono-space font such as a Courier or a terminal font. Maple output is formatted under the assumption that all characters have the same width.

The graphs that are produced in a Maple 3D plot window look best with a black background rather than with a background of a different color. If your system produces a background other than black for your 3D graphs, then you can use the following resource setting to specify a black background:

```
Maple3D*background: black
```

Adding the resource specification lines shown above to a file such as .Xdefaults does not change any of the resource values until the next time that you start using X, probably the next time that you log on. You can, however, force these new specifications to take effect right away if you use the xrdb program to read your .Xdefaults file:

```
xrdb -merge .Xdefaults
```

B.15   Tips

When there are many windows displayed on your screen, it is often useful to reduce the visual clutter by turning some of the windows into icons. The method for doing this is dependent upon the window manager that you use. When a window is reduced to a small icon on the screen, no information is lost and the icon can be expanded back to the window. The small size of the icon permits you to move them to a convenient corner of the screen out of the way of important windows. On-line documentation windows or graphics windows can be reduced to icons if you don't wish to see them for a while but you would like them again later.

As the number of windows created by Maple becomes large, you may notice your system responding more sluggishly. Each window does require some amount of system memory to maintain and reserving space for many windows can degrade the performance of your system. Whether the number of windows that you can have before you notice a change in the system's response is 10 or 50 depends on the speed of your machine and the amount of memory available. If you do have many windows or icons on the screen, you may find that closing or removing some of them will improve the performance of your machine.

If you will be showing the results of your work on the screen to others, it will be worth your while to experiment with using larger, and hence more legible, fonts. A bold Courier font in 18 or 24 point size works well for demonstrations. If you wish to switch fonts for the Maple worksheet window without changing your .Xdefaults file, you can type the following sequence of lines:

```
xrdb -merge Maple*Font: -adobe-courier-bold-r-normal--18-180-75-75-m-110-iso8859-1
control-D
```

The xrdb program is used to read the resource specification from the standard input file, which is the terminal in this case. Any number of lines can be read by xrdb and the control-D key combination is used to indicate the end of the input file. Note the long name of this font. If you type a name such as this one, you would be prone to entering it incorrectly. It's much easier to use xlsfonts to print a list of available fonts in your terminal window, use text selection to copy it, and then use the middle mouse button to paste this name after you have typed Maple*Font: .

B.16   Troubleshooting

When you start Maple and you see a message similar to "Cannot open display", then the X user interface program for Maple could not create a window on the display terminal which is given by your Unix environment variable DISPLAY. If this happens, then use the following command within your Unix shell program to see the current value of DISPLAY.

```
echo $DISPLAY
```

If DISPLAY has not been set properly for your display terminal, then you need to do so.

In addition, if you see a message similar to "Client is not authorized to connect to Server", then the system on which Maple is running is not allowed to create a window on your display terminal. To correct this, you should issue the following Unix command on your local display system:

```
xhost + RemoteMachine
```

where `RemoteMachine` is the name of the system on which Maple is running. This command lets programs running on the remote machine create windows on your local display system.

B.17   Information for Xperts

The Maple worksheet window, which has the widget name of Maple, is an Athena Text widget and thus understands all the Core resource names and classes as well as the Text resource names and class. In addition, it uses the private resources listed in Table A.1.

The widget names for the on-line documentation window, the 2D plot window, and the 3D plot window are maplehelp, maple2dplot, and Maple3D respectively.

The resources for Maple 3D plot windows are given in Table A.2. In addition, the command buttons which appear along the top in the menu bar are Command widgets. Examples of resource definitions which can be specified for these widgets are given below:

```
Maple3D*foreground: black
Maple3D*background: white
Maple3D*font: 9x15bold
Maple3D*titlecolor: blue
Maple3D*Command*borderwidth: 3
Maple3D.iconName: Maple 3D Plot
Maple3D.title: Maple 3D Plot
Maple3D.Translations: #override              \n\
    <BtnDown>: MouseTracking()               \n\
    <Btn1Motion>: MouseTracking()            \n\
```

```
        <Btn2Motion>: MouseTracking()                \n\
        <Btn3Motion>: MouseTracking()                \n\
        <BtnUp>: MouseTracking()                     \n\
        <Key>P: plot-action()                        \n\
        <Key>Q: quit-action()
Maple3D.anglebutton.Translations: #override \n\
        <Enter>: highlight()                         \n\
        <BtnDown>: set() notify() unset()            \n\
        <Leave>: unhighlight()";
```

Name	Default value	Explanation
Maple.inBufferSize	5	Size of the input buffer in Kbytes. This is used by the text widget to collect input for evaluation by the Maple kernel.
Maple.indentMode	on	If this mode is on, Maple indents one tab width after the "metaReturn" key is pressed.
Maple.replaceMode	on	If this mode is on, Maple will replace previous output in a cell with new results.
Maple.returnKeySubmits	on	If this is on, the "Return" key submits the current text selection to Maple for evaluation. If it is off, then "metaReturn" (or "Enter") has this role.
Maple.runDetached	true	If this is true, then Maple will run as a separate process in the background when started.
Maple.separatorLines	on	If this is on, Maple generates separator lines after each output expression.
Maple.tabSize	4	Width of a tab.
Maple.txtBufferSize	50	Size of the text window buffer in Kbytes. When this buffer is full, Maple will discard the top 1/3 of the buffer.

TABLE B.1. Private resources used by Maple

Resource	Class	Default value	Explanation
titlecolor	Titlecolor	White	Color of title text
titlefont	Titlefont	XtDefaultFont	Font for title text
foreground	Foreground	White	
background	Background	Black	

TABLE B.2. Resources for Maple 3D plot windows

Appendix C
Maple under DOS

C.1 Introduction

This appendix describes user interface features of the DOS version of Maple V. It is assumed that Maple V is correctly installed and running on your PC. To install Maple on your system, refer to the document titled *Installing Maple V on a DOS System*.

The DOS version of Maple V offers a user interface packed with features designed to increase productivity. Here's a list of some of these features.

- A graphics display driver for two and three dimensional plots has been added. The three dimensional graphics display driver has a menu which allows plotting options to be changed on the fly. By using the menu you can replot with different style options and rotation angles until you get the plot just right - then ask for a printout!

- Maple can now be driven from a menu system. Common operations such as loading library packages and simplifying expressions can be done from the Maple menu. The Maple menu is fully user definable. An easy to use on-line facility allows menu items to be added, removed and modified.

- A full featured text editor named MAPLEDIT for editing Maple expressions and files is included. MAPLEDIT can be called from within Maple allowing interactive development of Maple procedures.

- A log of the current session is automatically kept every time Maple is run. By pressing a single key, you can enter the Maple browser and review an up-to-date copy of the session log. This copy of the session log can be edited using MAPLEDIT. Before exiting the browser, a block of lines can be selected to be read back into the Maple session for re-execution.

- An input/output capture mode has been implemented. Using this facility, you can tell Maple when to start and stop saving a copy of the session input and/or output to a file.

- A command line editor lets you make changes to the line you are typing, and recall the last one hundred previously typed lines for editing and re-execution. A search facility lets you find a specific line by typing in a prefix of that line.

These features have been designed to provide flexibility as well as ease of use. For example, a more advanced user could set up a Maple menu which made it easy for Maple beginners to get started on a particular topic. Also, an effort has been made to be consistent throughout the user interface - thus minimizing the number of new commands you need to learn. For example, the same browser is used for choosing a help topic, displaying a copy of a help file, and reviewing a copy of the Maple session log; the same editor is used for editing Maple expressions, Maple files, copies of the session log, and copies of Maple help files.

C.2 Using Maple V

To start Maple V, type MAPLE at the DOS prompt. After a short pause while the program loads, you will be in the Maple session. You can now begin typing Maple commands. A good method of seeing Maple in action is to load help files into the browser and select help file command line examples to be read into the Maple session for execution. Section C.2.6 *Accessing Maple Help* explains how to do this.

C.2.1 Exiting Maple

To exit Maple, press the F3 key. A window will appear asking you to confirm that you wish to exit. Press Y to confirm. You can also use Ctrl-C to interrupt the current computation and exit Maple.

To interrupt Maple during a long computation, press the Ctrl, Alt, and left Shift keys simultaneously. The computation will be interrupted and the Maple prompt will appear in a few seconds.

C.2.2 The Status Line

The bottom line of the screen is reserved for use as a status line to display useful information about the current state of Maple. Since the same keys (i.e. functions keys) may have different actions depending on the mode Maple is in, the status line changes to remind you of some of the editing and function keys you may use in the current mode. For example:

```
F1-Help  F2-Find  F3-Quit  F4-DOS  F5-Review  F10-Menu   (  )  256k  17sec
```

While in normal session mode (right after you start Maple), the status line indicates the total number of bytes allocated by Maple as well as the number of seconds of CPU time used. To indicate when input/output capture mode has been turned on, an I and/or O will appear within the parentheses on the status line. Input/output capture mode is explained in Section C.2.9 *Input/Output Capture Mode*.

C.2.3 The Command Line Editor

When you are entering a Maple statement or command, you can use several of the keys on the keyboard to edit what you are typing. The ← and → keys can be used to move the cursor left or right one character at a time, without erasing any characters. The **Home** and **End** keys move the cursor to the beginning or end of the line respectively. The **Backspace** key moves the cursor left and deletes the character there. The **Del** key deletes the character at the cursor position. The **Ins** key turns insertion mode on or off; insertion mode is initially off. Pressing **Esc** while the control key is pressed will erase the entire line typed so far.

Sometimes, you will want to edit a line that you typed earlier. This is most common when you want to make a small change and see what effect it has, or you wish to correct a typing error that you didn't notice until after pressing **Enter**. Maple maintains a list of the last one hundred lines that you typed in. The ↑ and ↓ cursor keys can be used to move through this list. If you want to find a particular line quickly, you can type the first few characters of the line you wish to find and then press the F2 key. Maple will search backwards through the list for the most recent line that begins with what you typed. If no such line is found, the line you typed remains unchanged and the cursor moves to the beginning of the line.

C.2.4 Expression Editing

The previous section described how you can edit lines that you typed earlier. Using the Maple procedure `xed()` you can edit any valid Maple expression with with the Maple text editor (MAPLEDIT). The procedure `xed()` takes a single parameter. The parameter can be either an unevaluated name or an expression.

The parameter is evaluated and converted into line-printed form (the form in which you would type an expression). The MAPLEDIT text editor is then invoked to let you edit the expression:

```
▽
┌─┐
│f│ :=
└─┘
proc(x)
        if x <= 1
        then    1
        else    x*f(x-1)
        fi
end
;
△

  MAPLE.        1/8    1    4%    187K    ·    F1-Help  F2-Save  F3-Quit
```

The text editor makes use of the various editing keys (←, →, ↓, ↑, **Home**, **Ins**, etc.). Pressing the F1 key while the text editor is active will display a list of all the keys that you can use to edit. The MAPLEDIT Reference card provided with your Maple V documentation is a copy of

MAPLEDIT's on- line help.

When you have finished editing the expression, press F2 to save the changes you made, and then F3 to exit the text editor and return to Maple. The expression will then be read back into the Maple session. If the parameter you passed to xed() was an unevaluated name, the expression will also be assigned to that name.

On-line help for the xed() procedure is available with the command ?xed. The help file for xed() contains examples which demonstrate the different methods of calling xed().

C.2.5 File Editing

Just as xed() lets you edit an expression, the procedures fed() and fred() will let you edit a file. Both of these procedures take a name as their single argument and invoke the text editor (described in the previous section) on the file specified by that name. If the name contains any punctuation characters, it must be enclosed in back-quotes (`). The name must be given in Maple format rather than DOS format. The difference between Maple format and DOS format are that Maple uses a forward slash (/) rather than a backslash (\) to separate directory names in the filename path. For compatibility with DOS, each path component should be at most eight characters long. DOS is also not case sensitive (the Maple filenames MyFile and MYFILE map to the same DOS file names).

Both fed() and fred() function the same way except that fred() will read the contents of the file into Maple after you have finished editing it. This makes fred() very useful during the development of Maple procedures; you can go back and forth between editing and debugging a procedure without having to exit Maple. On-line help for fred() and fed() are available by entering ?fred or ?fed at the Maple prompt.

C.2.6 Accessing Maple Help

The usual method of accessing help in Maple V is to use the ? syntax. At the Maple prompt, type ? and then the topic for which you want help. For example, typing ?fred will load a copy of the help file for the Maple procedure fred() into the browser. If you type something after the ? which Maple doesn't recognize, alternative topics will be suggested.

The second method of bringing up a help file is with the F1 key. Pressing F1 will display a list of all subjects for which there exist help files. One of the lines will be displayed in reverse video. Use the ↑, ↓, PgUp, PgDn, Home and End keys to highlight a particular topic. Another method of choosing a topic is to search for it using the F2 key. Pressing F2 brings up a prompt asking for a search string. After entering a search string, pressing Enter will jump to the first occurrence of the search string. You can also press a letter key to jump directly to the first topic beginning with that letter.

After choosing a help topic using either of the above methods, you will be in the browser:

```
┌─────────────────────────────────────────────────────────────────────┐
│ ┌───────────────────────────────────────────────────────────────┐   │
│ │ FUNCTION: int or Int - definite and indefinite integration     │   │
│ └───────────────────────────────────────────────────────────────┘   │
│ CALLING SEQUENCES:                                                    │
│        int(f,x);   int(f,x=a..b);   int(f,x=a..b,continuous)          │
│        Int(f,x);   Int(f,x=a..b);   Int(f,x=a..b,continuous)          │
│                                                                       │
│ PARAMETERS:                                                           │
│        f           - an algebraic expression or a procedure, the integrand │
│        x           - a name                                           │
│        a,b         - interval on which the integral is taken          │
│        continuous  - (optional) indication that f is continuous       │
│                                                                       │
│ SYNOPSIS:                                                             │
│ - The function int computes the indefinite or definite integral of f with respect to │
│   the variable x. The name integrate is a synonym for int.            │
│                                                                       │
│ - Indefinite integration is performed if the second argument x is a name. Note that │
│   no constant of integration appears in the result. Definite integration is │
│   performed if the second argument is of the form x=a..b where a and b are the │
│   end points of the interval of integration.                          │
│                                                                       │
│ - If Maple cannot find a closed form for the integral, the function call itself is │
│   returned. (The prettyprinter displays the int function using a stylized │
│ ┌───────────────────────────────────────────────────────────────────┐ │
│ ↑,↓,PgUp,PgDn,Home,End-Move  F2-Find  Ins,Del-Select  F5-Edit  ↵-Do  Esc-Quit │
└─────────────────────────────────────────────────────────────────────┘
```

Use the ↑, ↓, PgUp, PgDn, Home and End keys to move about the copy of the help file. When you are in the browser you can highlight a block of lines which can then be edited or read into the Maple session. To highlight a block of lines, go to the start of the block you want to highlight and press the Ins key. Now scroll down to the line at the end of the block and press Ins again. You should now have a block of highlighted lines. In case you made a mistake, pressing the Del key unhighlights the block. You can perform two operations on the highlighted block:

1. Press the Enter key to quit the browser and read the highlighted lines into the Maple session.

2. Press F5 to invoke the Maple text editor with the highlighted lines. Edit the contents of the text editor. If you want to replace the contents of the browser with the contents of the editor, press F2 (save) and F3 (quit) in succession. You will be returned to the browser with the new contents. Note that the entire contents of the browser are now highlighted - allowing you to quit the browser and read the browser's contents into the Maple session simply by pressing Enter. Pressing Esc will exit the browser without reading anything into Maple.

Note that the browser only contains a copy of the help file. Editing this copy will only change the contents of the browser for this invocation. The next time you access help for the same subject, the browser will contain the original help file.

C.2.7 Session Review Mode

Maple automatically saves the last one thousand lines of your Maple session in a session log. This session log can be saved to disk or viewed with the browser.

Pressing F5 from the Maple session invokes the browser with a copy of the session log. When you are in the browser, you can highlight a block of lines which can then be edited or read back into the Maple session. The contents of the browser are manipulated in exactly the same method as explained in Section C.2.6 Accessing Maple Help.

Again, note that the browser only contains a copy of the session log. Editing this copy only changes the contents of the browser for this invocation. The next time you invoke session review mode, the browser will contain a new copy of the up-to-date session log.

C.2.8 Using the Menu

For the beginner, the menu system provides a simple interface to some of Maple's key functions. For the more experienced user, the menu offers a method of customizing Maple - relieving the user of the need to retype commonly repeated or difficult-to-remember command sequences. Modifications to the menu are done from the menu from within the Maple session. Thus, it is no more difficult to create a custom menu configuration than it is to use the menu itself.

To activate the menu, press F10. A menu bar will appear at the top line of the screen with a list of items to choose:

Items are chosen either by highlighting the item with the cursor keys and pressing Enter, or by pressing the letter key corresponding to the highlighted letter in the desired item. Invoking one of the top-level menu items will usually cause a submenu to be displayed. For example:

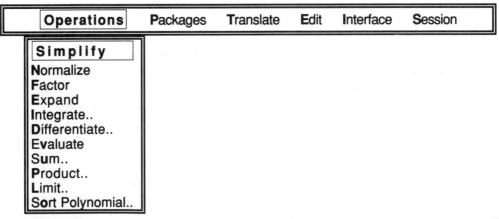

Items in submenus are chosen in exactly the same way as top-level items. Pressing Esc while a submenu is displayed will return you to the top-level menu. To exit the menu press Esc. The Session item in the top line menu bar is the only item that is common to all menu configurations. In other words, every menu will contain at least this item. All other items are definable by the user. The simplest menu would be one which contained only the Session item. Invoking the Session item will yield a submenu with the following items

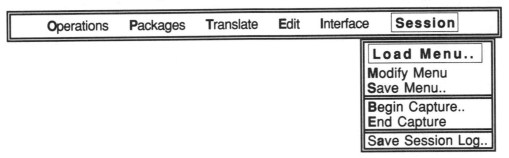

The items **Begin Capture**, **End Capture**, and **Save Session Log** are explained in Section C.2.9 Input/Output Capture Mode. The items **Load Menu**, **Save Menu** and **Modify Menu** are used to create new menu configurations. To load or save a menu choose the **Load Menu** or **Save Menu** item and enter the filename used to store the menu configuration. Menu configurations are stored to disk with the filename extension .MNU in the Maple library directory. In the case that the library directory is read only, the current directory will be used. Note that neither the path nor the extension should be entered as part of the filename since these are added automatically. It is a good idea to save the current menu configuration under a different name as a backup in case you plan to modify the menu.

When Maple is started, it loads in the default menu configuration. The default configuration is stored in the file MAPLE.MNU in the Maple library directory. If you have your own menu configuration that you would rather use as the default, save MAPLE.MNU under a different name (as a backup) and save your menu under the name MAPLE.MNU.

There are three levels involved in modifying the menu. The first stage is invoked by choosing the **Modify Menu** option in the **Session** submenu. The first level window is simply a window which allows you to select which top-level menu item to modify. The second level is activated after you make your selection. The second level window contains a list of the submenu items corresponding the the top-level item you selected in the level one window. After choosing a submenu item in level two, a large window will open, allowing you to modify the definition of the submenu item. The following gives a further explanation of each level.

LEVEL ONE

Invoking the **Modify Menu** option will open a window which has room for exactly sixteen entries. Each of the current top-level menu items except for the **Session** item will appear in one of the sixteen slots. Choose the top-level menu item that you wish to edit or choose an empty slot if you want to create a new top-level item. After you make your choice, a second window will open and you will be in level two.

LEVEL TWO

The level two window has room for exactly sixteen entries. Each of the current submenu items for the top-level menu item selected in level one will appear in one of the slots. At this point, there are two operations you may perform.

- Select Edit Menu Name to change the name of the top-level menu item you selected in level one. Setting the name to blank will delete that item from the menu.

- Select one of the submenu items to modify or choose a blank slot if you want to add a new item.

- Press Esc to return to level one.

After selecting a submenu item to modify, you will be in level three.

LEVEL THREE

A large window will open which lets you edit the definition of the submenu item chosen in level two. If you chose a blank slot in level two, you can now enter the definition for a new submenu item. The definition can be a maximum of sixteen lines long. As shown on the status line, the ↑, ↓, PgUp, PgDn, Home, End, Backspace, Ins, and Del keys are used to edit and move to different fields in the level three window. After you have completed modifying the item press F3 to accept the new definition. To delete the item, set the name of the item to blank and press F3.

There are two types of entries in the definition field lines: Maple commands which are to be read into the session when this item is chosen, and lines beginning with %. Lines beginning with % are used to obtain input from the user before the Maple commands are read into the session. All lines beginning with % (if any) must come before any Maple commands.

Lines beginning with %n (where n is a digit from 0 to 9) designate that the user should be asked for input when the menu item is invoked. The text following the %n is uses to prompt the user for input . When the user activates an item that uses the % notation in its definition, a dialog box opens and requests the user to enter expressions into each of the % fields. Then, all occurrences of % variables (not to be confused with % labels in Maple output; see ?interface) in the Maple code of the item definition are replaced with what the user typed.

Executing Maple code by invoking an item from the menu will not echo the Maple commands to the screen. To cause commands to be echoed, end the line of Maple code in the menu definition with a tilde (~). Only those lines ending with a tilde will be echoed. This feature is useful to prevent cluttering up the Maple session while still allowing informative lines to be echoed.

Here is an example of a menu entry from the default menu (MAPLE.MNU) supplied with the Maple distribution:

```
Command Name: [Sum..        ]
                      Maple Commands
[%1Find  sum  over                                          ]
[% 2  ranging  from                                         ]
[% 3              to                                        ]
[sum(",%1=%2..%3);~                                         ]
```

When this menu entry is invoked (by pressing F10, O, and ∪), the following dialog box appears:

```
Find sum over:  [                                           ]
   ranging from:  [                                         ]
            to:  [                                          ]
```

If you then fill in the fields like this,

```
Find sum over:  [i                                          ]
   ranging from:  [0                                        ]
            to:  [100                                       ]
```

the following Maple command will be entered into your session and executed (and echoed because of the ˜):

    sum(",i=0..100);

This will compute the sum of the previous result with the summation variable i ranging from 0 to 100.

C.2.9 Input/Output Capture Mode

At any time during the Maple session, you may specify that a copy of the subsequent input and/or output be saved to a file. This feature is accessed from the menu. To start capturing the input and/or output, choose the

Begin Capture item from the **Session** submenu. You will be asked to enter a file name and to specify (with "Y" or "N") whether the input and/or output is to be captured. After you have entered the requested information, capture mode will be turned on and you will be returned to the Maple session. To indicate that capture mode is currently active, the letters I and/or O will appear within parentheses on the status line depending on whether the input and/or output is being captured. To end capture mode simply choose the **End Capture** item from the **Session** submenu. After ending capture mode, the letters I and O will no longer appear on the status line.

To save the current session log to disk, choose the **Save Session** Log item from the **Session** submenu. After entering a valid DOS filename the session log will be saved to disk and you will be returned to the Maple session.

C.3   Manipulating Graphical Output

Plots are generated in Maple with the procedures `plot()` and `plot3d()`. For explanations and examples of how to use these functions, consult the on-line help with the commands `?plot` and `?plot3d`. The following sections explain how to display plots on the screen, obtain a printout of a plot and save a plot to a file.

C.3.1   Three Dimensional Graphics Display Driver

The three dimensional graphics display driver is used to display the results of a call to `plot3d()` on the screen. When a plot is displayed, a menu with nine items can be activated by pressing F10:

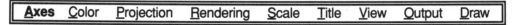

These items are explained below. An item in the top level menu or in a submenu is chosen either by highlighting the item with the cursor keys and pressing **Enter**, or by pressing the letter key which corresponds to the underlined letter of the desired item. To exit the display driver press **Esc**. On-line help is available with the command `?3Dgraphics`.

The Axes, Color, Projection, Rendering and Scale Items

Choosing any of these items opens a submenu of options. For example:

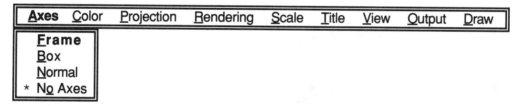

The currently selected option in a submenu is marked with an asterisk (*). Select an option by the method explained above – either by highlighting the option with the cursor keys and pressing **Enter**, or by pressing the letter key which corresponds to the underlined letter of the desired option. To exit a submenu press **Esc**. Except for the `Palette` item in the `Color` submenu, new choices will take effect only after the Draw item in the top level menu is chosen. Invoking the `Palette` item in the `Color` submenu will redisplay the plot using the next coloring scheme in a list of eight possible coloring schemes. For a complete list of the `plot3d` options available see `?plot3d[options]`.

The Title Item

Choosing this item opens a window in which a title for the plot can be entered (or edited if there was already a title):

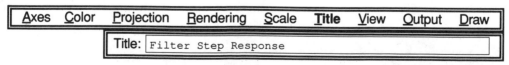

After the title has been entered, pressing Enter will cause the title to be displayed centered at the top of the screen. The title will be shown after the menu is exited.

The View Item

Choosing this item opens a window displaying a cube which matches the current values of the Phi and Theta viewpoint angles:

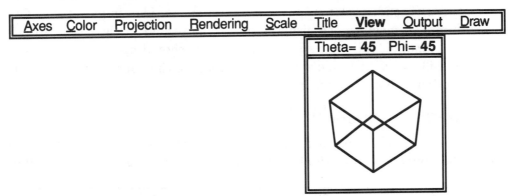

Pressing the cursor keys changes the Phi and Theta angles by five degrees per keypress. Pressing the cursor keys on the numeric keypad changes the Phi and Theta angles by a single degree per keypress. The cube rotates to match the new angles. Press Esc to exit. New choices for the angles will take effect after the Draw item in the top level menu is chosen.

The Output Item

Select this item when you want to send the plot to the printer or a file. After choosing this item, a window opens in which to enter the device type and the output file:

Axes	Color	Projection	Rendering	Scale	Title	View	**Output**	Draw

Device Type: `epson9`

Output File: `LPT1`

The plot output will be sent to a file unless the entry in the output file field is one of the device ports. See the Section C.3.3 Printing and Saving Graphic Output for a list of supported device types and device ports.

C.3.2 Two Dimensional Graphics Display Driver

This display driver is used to display the results of a call to the plot() function on the screen. When the plot is displayed, a menu can be activated by pressing F10. This menu has three items - Palette, Title and Output:

Palette	Title	Output

See Section C.3.1 Three Dimensional Graphics Display Driver for explanations of these items. To exit the display driver press Esc. On-line help for the two dimensional display driver is available with the command ?2Dgraphics.

C.3.3 Printing and Saving Graphic Output

There are two methods of sending plot output to a printer or file. The simplest method is to choose the Output item from the menu in the display driver. The second method is to specify that the plot() or plot3d() function call which generates the plot should send the plot output directly to the printer or to a file without first displaying the plot. See the appropriate section below for a step-by-step explanation of both these methods.

Supported Printer Device Types

Maple can generate graphic output for the following printer devices. You should use the abbreviations epson24, epson9, epson9hi, etc. when specifying the printer device to Maple.

epson24	EPSON compatible 24 pin dot matrix printers
epson9	EPSON compatible 9 pin dot matrix printers
epson9hi	EPSON compatible 9 pin dot matrix printer in hi-res mode
hp7470	HP 7470 plotter
hp7475	HP 7475 plotter
hp7550	HP 7550 plotter
hp7585	HP 7585 plotter
ibmpro	IBM Proprinter X24 dot matrix printer
ibmquiet	IBM Quietwriter dot matrix printer
laserjet	HP LaserJet printer
paintjet	HP PaintJet printer
postscript	PostScript printer
toshiba	Toshiba 24 pin dot matrix printer

Printing or Saving a Currently Displayed Plot

Printing the plot from within the display driver has the advantage of letting you rotate the plot and change any of the plot options before printing. Follow these steps to send the plot to the printer or to a file.

1. Generate the plot using plot() or plot3d().

2. Use the menu (activated by F10) to change the plot options, title, viewpoint etc. to your liking.

3. Redraw the plot by choosing the Draw item from the menu.

4. Repeat steps two and three until the plot is the way you want it.

5. Choose Output from the menu and enter one of the valid device types listed above and the name of the output file. If the entry in the output file field is one of {lpt1, lpt2, com1, com2} (see the list below) then output will be sent to the printer. If the entry in the output field is not one of {lpt1, lpt2, com1, com2} then the command sequence needed to print the plot on the the device type specified will be saved to disk under the filename specified.

lpt1	parallel port #1
lpt2	parallel port #2
com1	serial communications port #1
com2	serial communications port #2

6. While the output is being generated and sent to the specified file you have the option of cancelling the plot by pressing Esc.

Printing or Saving a Plot From the Maple Prompt

To print a plot directly, you need to specify the device type and output file to Maple. The following Maple code will print a plot on a 9 pin EPSON printer via parallel port number one. Substitute the appropriate device types and output file for your configuration.

```
interface( plotdevice = epson9 , plotoutput = lpt1 );
plot3d( x^2 + y^2 , x = -5..5, y = -1..3 );
interface( plotdevice = ibm );
```

The last command tells Maple to display future plots on the screen. For further information consult the on-line help for the interface() procedure using ?interface.

C.3.4 Using Maple Plots in Other Programs

Some software allows you to import graphical objects stored as PostScript or HPGL files. To create a PostScript or HPGL file containing a Maple plot simply specify the correct device type (eg. postscript) and an output file (eg. mypict.ps) when using one of methods of saving plots explained in Section C.3.3 Printing and Saving Graphic Output.

Index